Construction Practices for
Project Managers and
Superintendents

Construction Practices for Project Managers and Superintendents

W. J. Stillman

Reston Publishing Company, Reston, Virginia
A Prentice-Hall Company

Library of Congress Cataloging in Publication Data

Stillman, William James.
    Construction practices for project managers and
superintendents.

    Includes index.
    1. Building.   2. Construction industry—Management.
I. Title.
TH153.S83   690   77-24221
ISBN 0-87909-164-9

© 1978 by
Reston Publishing Company, Inc.
A Prentice-Hall Company
Reston, Virginia 22090

10   9   8   7   6   5   4   3   2

Printed in the United States of America.

This book is dedicated to Mechanics Institute, 20 West 44th Street, New York City—one of the oldest evening technical schools in America—and to the fine professional people who have been its instructors and administrators since its founding in 1820.

My good wishes are also given to The General Society of Mechanics and Tradesmen which, since its own founding in 1785, has served the construction industry, has founded Mechanics Institute as a free school, and has established one of the finest libraries in New York to serve Mechanics Institute and the community.

Contents

Preface

Almost every business or profession tends to create a sort of fraternity in which members or associates cooperate for everyone's gain. The construction industry is somewhat similar. However, people in the construction industry seem to go further; in addition to helping a fellow employee in the same company—they try to help all construction people who are seeking employment or are seeking instruction. Thus, it is not uncommon for a construction administrator who cannot hire an applicant to suggest another employer who might have an opening, or to suggest courses in instruction which might make the applicant more ready for employment with any construction company. In addition, most construction people try to look for an opening for any unemployed construction worker. Perhaps this is because the helper (in an industry that is noted for changes) may wish to feel that he can count on help at a later date from someone else or from the person he helped.

However, in the author's opinion, the help that one receives from other construction people is truly the help given by members of a very special fraternity; we try to help **all members** of our fraternity or those hoping to join our fraternity without regard for personal gain. This textbook is, then, another endeavor to assist young people who are first considering construction as a life's work or, more importantly, to assist those who have already chosen the construction industry for their life's work and now need further instruction so that they can, more efficiently, pursue this chosen profession. This textbook cannot be large enough to cover **all** facets one might encounter in construction. It does, however, cover most normal requirements and it does suggest reading of other texts on special subjects.

This text has been written with the hope that it will help its readers in what I feel is a wonderful and most fulfilling industry. During the writing of this text and the compilation of the illustrations I have been advised and helped by many construction men, engineers, architects, and leaders in the industry. I sincerely thank them for this assistance.

W. J. Stillman

Introduction

When our forefathers came to this country, a building construction project, from the design phase, to the foundations, the framing, the enclosing, the roofing, and all other appurtenances, was the effort of one man. This man designed the structure and, as the construction phase progressed, hired the necessary helpers and saw to the structure's completion himself.

As buildings became larger, and new innovations such as plumbing, central heating, and electric lighting were introduced, the building industry became one where specialist mechanics were employed. With further developments and larger projects, the building industry not only employed specialist mechanics, but also specialist contractors (or subcontractors), in addition to a General Contractor. Nevertheless, as the building industry has developed, it is still the responsibility of the General Contractor, the General Contractor's Project Manager, and the General Contractor's Project Superintendent to be fully familiar with all phases of the construction and the work of all subcontractors, so that the project will be correctly coordinated and the final product in full compliance with design requirements and the various codes that regulate the construction industry.

In most localities the building permit bears the name and signature of the Project Superintendent. He is accountable to the designers and the Building Department. Thus, he must have knowledge of all phases of an industry that has become an industry of specialists. He must be able to judge whether these specialties are being accomplished in a good and workmanlike manner and to oversee the project so that it is accomplished in this manner. One should not infer from this statement that the Project Superintendent must be a structural engineer, carpenter, bricklayer, plumber, electrician, steamfitter, *and* roofer; rather he should have a good working knowledge of all the trades employed on the project so that he can be aware of the design requirements and be sure that they are met.

Therefore, if the individuals that supervise a construction project must have these abilities, a text for their instruction must address itself to the many phases and divisions of a construction project. Our presentation of the different facets of construction projects will be set forth in the general order in which they will be confronted in the field. That is, after a discussion of the usual organization system of a construction company, we shall approach the projects with discussions on layout, foundations, superstructure, enclosure, roofing, and general completion, and in that order. A construction project is a complex accomplishment, and the subjects will be presented and amplified as required.

As construction projects have increased in size and as the costs of projects have increased, the method in which they are accomplished has changed. Originally, the project was the work of one man. Later, a General Contractor completed the project using his own forces for general construction and subcontractors for specialties such as tile, terrazzo, plaster, paint, roofing, electrical work, plumbing, and heating. More recently, the General Contractor has subcontracted **all** the work, using his own forces for jobs such as supervision, layout, protection barriers, and cleanup. Finally, many larger projects are now being accomplished under the direction of project management organizations who contract all the individual subdivisions to any number of prime contractors. Regardless of the system used, the methods of supervision are, essentially, similar. And this supervision must be accomplished by men who are fully knowledgeable about all the facets of construction.

The general methods in which a project is accomplished have not changed and, thus, the requirements for its supervisors and the knowledge they should have has not changed essentially. This text will address itself to the needs of any construction project, regardless of the contractual manner in which it is produced. The reader will be able to use what he has learned on any project, regardless of contractual systems.

Many of the subjects we shall discuss, if covered fully, could be the subject of an entire text. In fact, there are excellent texts on many of these subjects. However, this text is written to give the reader guidance on most of the problems he will encounter on a construction project. The reader must decide where additional study is needed; where we can, we shall suggest additional references.

Chapter 1

Organization of a Construction Company

The chain of command within the executive departments of one construction company may differ from that of another, but the duties performed by the executives are similar. However, the chain of command in the operations end of the company, from the Project Manager down to the last and lowest field clerk, is almost exactly **similar** in most construction organizations. True, one company may give the title of "Office Engineer" to an Assistant Construction Manager while another company gives him the title of "Job Runner." Usually, however, most companies use the same titles, and all companies use these people for the same duties.

Figure 1–1 is an organization chart that reflects the working systems used by most General Contractors. As we have previously noted, the systems of the executive departments may vary in different construction organizations. Therefore, it follows that different texts and different teachers may show charts which list the position of Chairman of the Board, President, Vice-President in charge of New Business, Vice-President in charge of Main Office Operations, Vice-President in charge of Field Operations, and similar titles. Other charts will list General Superintendents and Chief Engineers. However, regardless of the titles given and the manner in which this executive work is parceled out, in different companies, this is **management** work. Inasmuch as we are primarily concerned with the *field* operations of a construction company and the manner in which the construction process is regulated in the field, the organization chart in this text is simplified by listing this executive personnel as "**MANAGEMENT.**" Thereafter, and progressively down the chart, we list individual functions and individual titles. On larger projects there will be assistants to some of these people to handle a portion of their work.

Basically, there would be no need for Project Managers, Superintendents, and field organizations were it not for the "new business" endeavors of the management personnel. To bid a project with the hope of receiving a contract to build it, office personnel such as estimators and purchasing agents must take off the quantities of materials and equipment to be built into the project and must calculate a total cost (which includes a reasonable profit) to present to the owner and his designers. This is the company's bid.

3

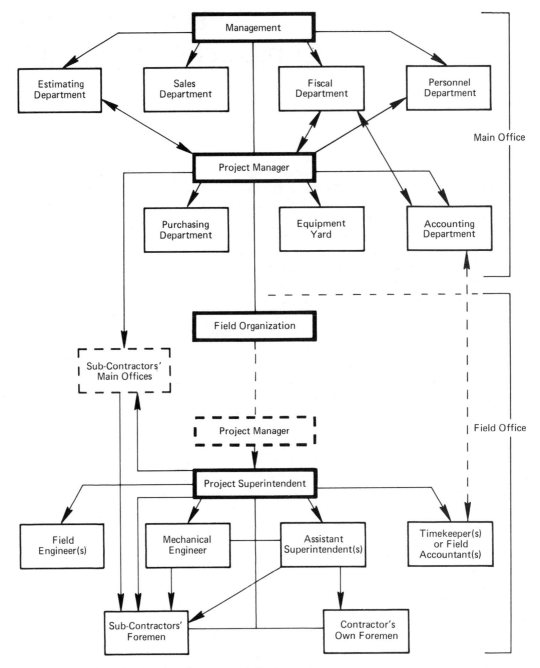

Figure 1–1. Organization chart for a construction company.

4

Thereafter, if the construction company is successful in receiving the contract to build the project, these same estimators will refine the base estimate and prepare a most accurate and more detailed take-off listing (which would call out special shapes, for instance) for the guidance of the men who will build the project; the purchasing agents will start taking bids from specialty contractors who wish to be subcontractors on the project. The purchasing agents will also get figures from suppliers and manufacturers for any of the material and equipment that the General Contractor intends to install himself.

When there is sufficient time a General Contractor will take bids from every trade, including a number of specialty contractors or subcontractors, and come up with a relatively exact estimate for the entire project. However, if time is short, it is a normal practice for a General Contractor or Construction Management organization to make take-offs or get precise bids on specialty trades (such as plumbing or electrical work), and to use unit-price costs of recent projects on other items or trades. Later, when the General Contractor or Construction Management organization has been awarded the entire project, it will take exact bids for all trades from several subcontractors in each trade, and will endeavor to negotiate a subcontract or buy out a subcontract at a price near to the original rougher estimate used in the original bid to the owner. If the (final) subcontract price is slightly lower than the original (rough) estimate, it will show as an under-run on the spread sheet (i.e., the final buy-out tabulation). If the subcontract price is higher than anticipated in the original bid, it will show as an overrun.

Whereas the estimates for subcontracts and purchases were originally made (for bidding purposes) under the direction of Management, the more refined and exact estimates and subcontractors' biddings are made under the direction of the Project Manager now that his company has a contract to build the project. The Project Manager will meet with specialty contractors desiring to bid this project, and (often after consultation with Management) will decide which specialty contractors will receive the subcontracts once bids from the desired subcontractors are tabulated. At the same time, and under the direction of the Project Manager, the Purchasing Agent will take bids and give out contracts (or purchase orders) for materials and equipment to be built in or supplied by the General Contractor. These functions—the purchasing of equipment to be built in or installed by the General Contractor and the contractual negotiations made with successful subcontractor—are usually called **buying out** a project. The reader would do well to remember this term, as we shall discuss in later chapters whether a certain item was "bought."

Copies of all the estimates and copies of all subcontracts, as well as copies of all equipment and material orders, will be given to the Field Office for file and for the knowledge and guidance of the Project Superintendent and his close assistants. Most often, a copy of the primary contract with the owner is given to the Field Office, along with copies of the plans and specifications for the project. However, in most cases, this guidance copy of the primary contract(s) will have the contract monies obliterated or masked out. There may be no reason for the people in the Field Office to know these facts, and the General Contractor will not wish these figures to get into the wrong hands. However, the *terms* of the contract must be known to the Field Office people.

1–1   Organization of a Main Office

As noted earlier and as shown in Figure 1–1, all main office executive personnel are listed as "Management." Thereafter follow an Estimating Department, Sales Department, Fiscal Department (usually headed by the Treasurer, who is a part of Management), and Personnel Department. In small companies these functions could be handled by five or less employees. In larger companies there would be a number of employees in each department and, most probably, each department would be headed by an officer of the company.

The Project Manager comes next on the chart. If he was available from the beginning, he may have had a part in the bidding of the project. Now, however, when he is actually assigned to the project, he will be in charge of the project and will deal with all departments in the company. He will buy the subcontracts and, with the help of the Purchasing Department, contract for the supply of items such as reinforcing steel, concrete inserts, miscellaneous iron, and doors and door bucks if they are to be installed by his company's own forces. He will keep the Superintendent continually informed and, as previously noted, will have copies of all contracts and purchase orders sent to the Field Office for its files.

You will note in Figure 1–1 that the Project Manager is shown as a part of both the main office personnel and is also shown as a member of the field office personnel. The actual location of his office depends on the size of the project. If the project is small, he will handle it from the main office and will, undoubtedly, handle other small projects. In these instances he will be in frequent contact with the Project Superintendent, who will run the field office and direct all field operations. However, if the project is large and requires the full time of a Project Manager, he may move his office into the field. At the discretion of the Project Manager, certain members of the Purchasing and the Estimating Departments may move into the field with him.

1–2   Organization of a Field Office

Whether the Project Manager has his office at the main office or at the field office, the Project Superintendent is in charge of all field operations and all personnel involved in these field operations. Not only will he give orders to all personnel on the field office payroll, but he will also be in charge of the subcontractors and the work of their forces. Whereas the Project Superintendent is shown as a part of field organization and under the Project Manager, the Project Superintendent is responsible to his employer, to the Department of Buildings, and to the owner or client to insure that the project is constructed in full compliance with the contract documents and with applicable building codes. Thus, even though he falls under the direction of the Project Manager, he is in charge of the actual construction program. Even on a relatively small project, he will have the assistance of a Field Engineer, who will lay out the building lines and set the grades (these operations are called "line-and-grade" operations); he will also have the assistance of a timekeeper, who will keep track of hours worked by company employees and the manner in which these hours are charged to different portions of the contract. In addition to field office staff, he will be in charge of foremen on his company's payroll and will give direction to the superintendents or foremen of his subcontractors. In this regard, it is important that the Project Superintendent have the ability to lead his subcontractors' field personnel so that the project's schedule is closely followed, rather than having to *push* these people. He must have a knowledge of the construction industry so that he is always aware of the quality of every trade's work, and he must have the ability to get the work done with cooperation rather than force whenever this is possible. The leadership of a good superintendent is often the key to a successful and "happy" project.

If the project is large, the Project Superintendent will have Assistant Superintendents and, perhaps, the assistance of a Mechanical Engineer, who will help him coordinate the work of the mechanical/electrical trades and will supervise the work of these trades under his direction. On a larger project, his Field Engineer will have assistants. With the addition of more personnel on the company payroll, there will be a need for more people in the Timekeeping office and at least one secretary for field office typing.

The Field Engineer is responsible for the layout of the project and setting the grades for the project. He is responsible to and takes his orders from the Superintendent. However, in actual practice, he usually runs his own work once he and the Superintendent have had an initial conference and have discussed the needs and schedule for the project. The Field Engineer (and his assistants on a larger project) will carefully watch the project so that line and grade are properly set for the construction of concrete forms, and that these forms are checked for location and have pour-grades prior to the pouring of concrete. The Field Engineer will check structural steel with a surveyor's transit to ensure that the steel is plumb, and he will check the level of the base plates (or billet plates) to be sure that the structural steel is at the correct elevation before he or the Project Superintendent give the steel erector permission to install final bolts or rivets. He will give line and grade to mechanical trades.

Assistant Superintendents tend to be the "legs and eyes" of the Project Superintendent. Most certainly must the Superintendent walk his project daily and keep his eyes open. But, as the size of the project increases, a Superintendent is called upon to fulfill more office duties and contact more

people by telephone. In these cases he needs assistants who have the knowledge to supervise the work, give orders under his authority, and to run the
many, many errands that confront a construction man daily. These Assistant
Superintendents will confer with foremen and with Field Superintendents
of subcontractors, and will make judgments and give orders for the Project
Superintendent where necessary. And, they will confer with the Superintendent often so that he is constantly advised.

On larger projects there is often a need for someone to supervise the
work of the mechanical/electrical trades and to help them to keep the
overall schedule for the entire project. On such projects the General Contractor often utilizes the services of a Mechanical Engineer. This person
should have sufficient formal education or field experience so that he can
help these trades when they have conflicts with the work of other contractors or with the structure itself. The Mechanical Engineer must always remember that he works for the General Contractor. Because he spends the
majority of his time with the subcontractors, there is a tendency for him to
become loyal to the mechanical/electrical subcontractors in opposition to
other contractors on the project. A good project Mechanical Engineer must
try to be friendly with all subcontractors, as must the Project Superintendent and other members of the staff. Nevertheless, primary loyalty must be
to the employer and to the project. Also, the project Mechanical Engineer
must have enough knowledge of structural design (in addition to his knowledge of mechanical/electrical design) so that he will not allow the trades
under his direction to put loadings onto a structural member at a point
where it cannot adequately support the load, and so that he will not allow
the trades under his direction to penetrate a structural member when the
structural design has not made provisions for such penetration.

Every project has a timekeeper. This person compiles the hours from the
time-cards passed in each afternoon by company foremen and sets forth
hours on the payroll. The weekly wages of "overhead" personnel, such as the
Superintendent and the Field Office staff, go on this payroll also. In addition, many construction organizations require that the foremen break out
the daily hours of each man on their cards to indicate the different work
being done by each man. In these cases the timekeeper keeps a weekly
summary of these breakdown figures. At the end of each week, the timekeeper gives a report on the number of hours the laborers spent against such
items as hand excavation, common brick, face brick, and how many hours
bricklayers spent against such items as common-brick and face-brick installation. These costs are sent to the main office along with a report from the
Field Engineer(s) as to the amount (quantities) of hand excavation, common-brick installation, face-brick installation, and the like, accomplished, so
that a unit cost can be figured. A cost report is prepared from these figures
so that the Project Manager and the Superintendent can know if their project is achieving the "bid" unit prices; the project summary of such unit
prices may also be used for future bidding. If these weekly cost reports are
compiled in the Field Office and if other work, such as payment of the
project's bills, is handled by the Field Office, then the timekeeping force will
be augmented by additional field accountants.

If the reader has been following the organization chart in Figure 1–1, he

will notice that we have not discussed one item. On the left, the heading Subcontractors' Main Offices is shown in a dashed box to indicate that it is not a portion of either the General Contractor's main office or field office. However, as soon as a General Contractor (or a construction organization) has subcontracts with specialty contractors, there will be negotiations between the Project Manager and these subcontractors. And, as soon as the Project Superintendent comes into the picture, he also will be dealing with the subcontractors' main offices to be sure that the men and equipment for which these subcontractors are responsible are on the project when they are required. Thus, in this regard, as soon as the subcontractors have a foreman and men on the project, their field organizations come under the orders of the Project Superintendent and his assistants.

1–3   Implementation of the Project

We have now discussed briefly each phase of a construction organization's main office and field office. In Chapters 4 and 5, we shall start to discuss the actual implementation of this staff and the manner in which the project is started.

However, nothing in any business "just happens" without long and careful planning. In the construction business planning will embrace such items as the construction schedule and the types of machinery, excavators, and hoist systems that will be used on the project. This planning was superficially covered in the original thinking that was necessary to be able to prepare the bid. However, now that the construction organization has a contract to produce the project, the Project Superintendent must make concrete plans on how he wishes to run his project and how he will, actually, set up the project. If his decisions differ from his company's original thinking, he will discuss his own ideas with management and the Project Manager so that final decisions can be made.

In addition, the Project Superintendent will make arrangements with the purchasing department to order equipment he wishes to buy and to rent other equipment. He will decide if the preliminary layout of the building(s) is to be made by his own Field Engineer and/or by an independent surveyor. This "and/or" is very important and will be discussed in Chapter 4. However, initial layout is often done by an independent surveyor for legal reasons. If the services of an independent surveyor are to be used, the Superintendent must arrange for this work well ahead of time. He will also check with the Project Manager and the Purchasing Agent to be sure that materials such as trans-mix concrete, reinforcing steel, masonry, door bucks, hardware, and other long-lead items have been ordered or contracted for. And, if the construction manager and the main office have not filled his Field Office staff with employees currently on his company's payroll, he will make arrangements with the personnel department so that unfilled positions will be filled when required.

Chapter 2

The Designer of
the Project

Anybody can design a building. If the reader doubts this, he can find scores of wives who (with or without the help of their husbands) have "designed" and made drawings for the family residence. In many cases these drawings were then given to an Architect who perfected them so that the desires and needs of the owners were met, and, more importantly, the design really worked! In some of the cases that were not referred to professional designers, many of the kitchens did not work out well, some of the desired closet space was not available, and other items such as stairs were awkward.

Industrial buildings could be designed in the same manner. The industrial engineers and managers of a manufacturing company know, better than anyone, how the assembly lines should be located and how certain departments should be placed in relation to other departments. Considering that these functions are engineering functions and, therefore, that the building is a product of engineering functions, perhaps these engineers should design their own factory buildings. Usually, however, **they do not.** Rather, they take their layout drawings to an Architect, who, after full study of their drawings and many conferences with the industrial engineers and company officials, makes a design that is in accordance with both manufacturing needs and good architectural practice.

What then is an Architect? A good Architect is a rare combination of an artist who can design a structure that is pleasing to the eye and structurally adequate, and also meets every need of its owner. Because the Architect's business depends on his ability to do the best design for his clients, he is continually updating his original knowledge of the construction industry by investigating new products and new building systems. This updating of knowledge consumes considerable time, because new systems are continually being introduced to the construction industry. Some are good, and some will not stand the test of time. Considerable evaluation is required. An Architect who has a good fundamental background and training, who has kept up with all the good new construction innovations, and who is himself a good businessman can usually produce the best building design for a company. It is for these reasons that most knowledgeable industrial companies do better to use a good architectural firm than they do by using in-house designers.

Years ago the Architect designed the entire building. Now, however, most architectural firms take projects as "joint-ventures" with a Structural Engineering firm and a Mechanical/Electrical firm. Just as the purely architectural processes are changing each year, there are new innovations in structural and mechanical/electrical systems. Thus these engineering firms must keep up with the times. Also, certain Structural Engineers may have developed a fine reputation for a special type of construction, and certain Mechanical Engineers have developed a fine reputation for specially controlled heating or mechanical systems. Therefore, an Architect may use different engineering firms as consultants or joint venturers on different projects. He can use the best engineering firms for the special needs of his client and the special needs of the structural system he has chosen for his client. Thus his designs (and designers) can be as flexible as each project requires.

2–1   Mechanics of the Design System

After the Architect has determined the general needs of a client, he and his consulting engineers sit down in conference with the client's people and they often send design teams into the client's offices and factory to make studies of the particular requirements for the new project. They will then go back to their offices and make initial designs which will be presented to the client for review by him and by his company's specialists. Further consultations between the designer's and the client's specialists will take place as the designs and final drawings are produced.

2–2   Final Drawings and Specifications

Even as the architectural, structural, mechanical, and electrical design drawings are being produced, specifications (i.e., the verbal description of the requirements of each material or operation involved in the design) are being written. The specifications should be written in close harmony with the needs of the drawings and each detail of the drawings, so that there can be no question as to what the designer intends and what the contractor will have to accomplish to complete his contract to the satisfaction of the owner and the designer. When the drawings are finally completed, the specifications written, and both sets of documents have been approved by the client, the Architect will present the drawings to examiners in the Building Department of the community where the project is to be erected. When it has received the approval of the local Building Department, the drawings will be presented (on a loan basis) to several contractors for bidding purposes.

2–3 The Fast-Track System

Quite often these days, the step-by-step design process is "detoured" to what is termed a "fast-track" system. In this system, a General Contractor or Project Managing organization works along with the designers and, when both are agreed that the design is best, takes bids, awards contracts, and starts the actual construction before the design for other portions of the project are completed. Thus the excavation and foundation work is commenced before final design drawings for other facets of the project are completed. This system has the advantage that construction may begin sooner and, thus, give the client a new and (usually) more efficient facility sooner. It has the disadvantage that very often it presents the owner with "extras"[1] that might not be encountered if the complete design were available before any construction began. The current economics of the area and the economics of the client's needs must be balanced with each other to decide if the fast-track system or the conventional design-first, bid-next system will be used.

2–4 The Architect's Authority

We must always remember that *the Architect is the representative of the owner* and, also, that it is the Architect's name and license-seal that appear on each drawing. He has designed the project and he has assured the authorities that the project, if built in strict accordance with designs and specifications, will achieve all requirements of the local building code. In many communities he is required to so attest in writing before the building can receive a Certificate of Occupancy. This "CO" must be obtained before the client can take beneficial occupancy of any building. Thus the Architect has the authority to review and approve subcontractors before they can receive contracts and the authority (and he is required) to inspect construction as it proceeds. Thus, regardless of whether the project is designed and built under the conventional-design first and bid-next system or by the "fast-track" system, the owner, Architect, and builder must work in complete accord with each other. The contractor must remember the Architect's position in the legal chain of command, and must make every effort to insure that the project is built and completed to the satisfaction of the Architect.

There are many other duties that the Architect and his design consultants must perform. However, so that we may proceed with duties of the builder, the complete duties and functions of the Architect will be postponed and more fully discussed in Chapter 28.

---

[1] An extra is payment for work not covered by the contract.

Chapter 3

Types of Contracts

3–1 The Contract Documents

Before we can discuss the different types of contracts under which construction projects are achieved, we must have a knowledge of what are termed "The Contract Documents." Obviously, the formal contract between the owner and the General Contractor or construction organization is a portion of the contract documents. However, in this regard, the formal contract states the terms and monies for the construction of a project in conformance with the design drawings and specifications. Thus the formal contract, the design drawings, and the specifications are always interdependent and are always considered as the contract documents. However, an addendum (if more than one, called addenda) *may* be a part of the "contract documents," whereas the design drawings, the specifications, and the formal contract are (normally) considered the "contract documents," sometimes the "addendum" or a number of addenda become a part of the contract documents.

For example during the final phases of design, the Architect must make changes in certain designs or in portions of the specifications, and he finds it more efficient to issue additional drawings or specifications rather than make changes in the original documents. These additional plans and/or specifications are issued in what is called an **addendum.** Because there may be more than one, each addendum is numbered. These become part of the contract documents.

After considering the contract documents, one must consider how these documents will be covered by a contract with a construction organization. The contracting systems vary, and we will discuss the different types in the following sections.

3–2 Lump-Sum Contracts

From the very beginning of the formal construction industry, the lump-sum type of contract has been the type initially considered. Under this system the contractor, after making careful estimates of the amounts and

unit costs of work or material shown on the contract drawings (i.e., the
design drawings that will be a part of the "contract documents"), makes a
proposal which states that he will build the entire project for a certain
amount of money. This amount of money will not change regardless of
future additional costs due to his miscalculation, price rises in materials or
wages during the term of the project, or other items such as cost increase of
subcontracts. Under this type of contract the owner knows, from the very
beginning, the total cost of his new project. This agreed price will not
change unless additions are made to the contract drawings. Parenthetically,
however, we must state that under the lump-sum and other types of contracts which we will discuss in this chapter, the contractor is protected from
responsibility of what is termed "Acts of God." That is, he will not be held
accountable for circumstances or conditions he could not anticipate when he
made his estimates and submitted his bid.

For example, at the time a contractor makes his bid he knows he must
anticipate a certain amount of precipitation, and he knows if current labor
agreements are subject to change during the life of the contract. Such items
are considered to be his responsibility. However, if the area receives considerably more precipitation than the average for the area as shown by the
weather bureau of the Department of Commerce, or if there are hurricanes,
tornadoes, or other extreme weather conditions that one would not antici-

Helicopters are great aids in modern construction. (*Courtesy of Bell Helicopter, Division of Textron*)

pate, he is not held accountable for the extra cost or delays incurred. If there are community-wide strikes (which delay the project) that were not caused by his action, he will not be held accountable for extra cost or delays caused by these strikes. However, if action on his particular project was the cause of a strike on his project, he may be held accountable, and he may not receive extra payment for losses so incurred.

Every type of contract should clearly note the extent of its coverage. Thus, if new items desired by the owner or if incorrectly shown items already on the contract drawings necessitate changes in the contract drawings (usually noted as "revisions"), the contractor will claim an "extra." If the contractor suffers losses due to abnormal weather (which should be checked for the entire life of the project in order to protect both the owner and the contractor), he will eventually be due an extra. Extras will be discussed at the end of this chapter.

3–3   Cost-Plus Fixed-Fee Contracts

There are often reasons that a contractor cannot or will not risk a lump-sum type of contract. The size of the project or local labor conditions may make the contractor unwilling to take such a chance. The drawings may not be finalized but the owner may wish to start construction. Or the progress of the entire project may have been originally calculated on a "fast-track" basis, which we discussed in Chapter 2.

Under these situations the contractor may agree to keep precise records of all money he spends for labor and material on his own portion of the contract and for the cost of subcontracts and bill the owner for these costs *plus* a previously agreed upon *fixed* fee for the entire project. Of course, in such systems as the "fast-track" design system, this fee will be based on a fixed percentage of the actual, final cost rather than a total fee for an amount the contractor could not accurately estimate at the inception of the project.

An often-voiced misapprehension is that under the "Cost-Plus" type of contract there is no way that the contractor can lose. Nothing could be further from actual fact! For example, if the fixed fee for the entire project is $200,000 and if the estimated life of the project is two years, then the contractor is anticipating a profit of $100,000 per year from that project. This $100,000 per year is earmarked to pay the rent on the main office, the salaries in the main office, and for a nominal profit for the owners or stockholders. But if delays cause the project to run for three years instead of the anticipated two, the annual funds received from the project are only approximately $67,000. The $33,000-per-year-difference could cause an actual loss to the company operation. A project that has delays which cause the contractor to tie up a good field office crew and a proportionate portion of his main office personnel for a considerably longer time than anticipated can most surely reflect actual (net) losses to the company. Also, if the project

had been completed on schedule, these same people could be placed on another project where they could be instrumental in future profits for the company.

We have considered the advantages and disadvantages of the Cost-Plus contract from the contractor's point of view. There are also aspects that must be considered by the owner and his Architect before entering into this type of contract. There are, most certainly, advantages in starting the project before the design drawings and specifications are completely finalized. One primary consideration is that the owner will have his new plant sooner and will be able to anticipate greater profits from his operations. However, if the design of the new facility is almost completed, he may do well to wait until he can expect a contractor to enter into a Lump-Sum contract.

For example, even if the contractor offers a Cost-Plus Fixed-Fee contract that includes an "upset sum" clause (i.e., a contract which stipulates that the contractor will guarantee that his costs, plus his fee, will at no time ever exceed a contract-stipulated top sum), there is a possibility that the owner could lose money. Although it might seem impossible, it happens. For example, if the contractor does not make every effort to keep the original costs as low as possible, or if he watches them only to the extent that they are allowed to rise to a point where total costs plus his fee do not exceed the "upset sum," his poor management costs the owner money. However, if the contractor's only incentive is to protect **his own** interests, he may not be trying to save the owner money. Why should he? He has no added incentive. The next contract type gives the incentive and is, perhaps, a contract more acceptable to both the owner and the contractor.

3–4 Cost-Plus Fixed-Fee with Additional Clause for Profit Incentive

The one hazard that the owner faces with the previous Cost-Plus contract is that the contractor may not make sufficient efforts to insure that the project is built as efficiently as possible. Of course, there is the possibility that his organization does not have the talent for the most efficient management of the project. More often, however, costs are allowed to build up until they endanger the contractor's profit margin. Please understand, no accusation of negligence is intended; but in today's manpower market, even the most honest of contractors can let this happen.

There is a type of Cost-Plus contract that makes it profitable for the contractor to make every effort to save money for the owner. This is a Cost-Plus contract which gives the contractor a share of the cost and the upset sum set forth in the original estimate. The contractor has advised the owner and his architect that he estimates that the total cost of the project will not exceed a certain sum and that his fee for this work will be X dollars. In addition, he stipulates that he will guarantee that the cost of the contract, plus his fee, will not exceed a certain amount. This is a part of the price that is recorded in the contract. However, in addition to the guaranteed Cost-Plus-Fixed-Fee, the contract stipulates that the contractor will receive a share (quite often 25 percent) of the savings between the original cost estimate (which is noted in the contract) and the actual (lower) cost. If the cost goes above the original estimate, the contractor bears this cost. This system gives notable incentive! If a project is to be built under some kind of Cost-Plus contract, this system may be the best answer.

3–5   Several Bases for the Award of Contracts

Federal law and legislation in most States mandate that public construction contracts be awarded to the lowest qualified bidder. Many U.S. cities and larger municipalities have passed similar legislation. Thus Federal and State contracts go to the lowest bidder; and even though a particular municipality does not have specific legislation that binds local authorities, contracts are usually given to the lowest bidder to avoid possible criticism from cost-conscious citizens. Quite often municipal executives may feel that another contractor could do a better job and that the price difference would be saved in the long run. Regardless, they rarely take chances with their own careers, and thus they too give the contracts to the lowest qualified bidder.

Contracts awarded by corporations are subject to a different set of rules. The basic rule is that the best interests of the corporation must be served at all times. The following are a few of the reasons that a contract could be awarded to a higher bidder:

1. One of the bidders has the ability to complete the project in a shorter time. This will bring a new, more profitable facility into use sooner, so the difference between this bidder and the lowest bidder is not the most important factor.

2. One of the bidders has had more experience in the corporation's type of construction.

3. One of the bidders has successfully completed a number of projects for the corporation, and the corporation's executives have knowledge of him and confidence in his work.

The executives of a corporation will seriously consider these aspects and others and will then make the decision. However, in making the award, these executives may take extra precautions to protect their company's interests, especially if one of the reasons for award to one of the contractors is early completion. Thus, if the executives have chosen a contractor with a higher bid because he has agreed that he can complete the project sooner

than the date originally anticipated, they will endeavor to protect the interests of their company by putting a "penalty clause" into the contract.

3–6   Penalty and Liquidated Damage Clauses

Penalty clauses have been inserted in contracts for decades. Basically, these clauses state that, if the contractor does not complete the project to the point that the owner may take *"beneficial occupancy"* by the end of the contract period, he will pay the owner a specified amount per day until he does complete the project. This amount is deducted from his final payments, actually. The original conception was fair. If an owner was building a facility to save future operation costs with the most efficient, new facility, then truly he lost money when he could not use the new facility as soon as had been promised. In such cases, his daily loss was deducted from the price paid to the contractor.

However, far too often the penalty-clause system was used to reduce the contract amount at times when the owner did *not* lose money by waiting for the completion of his project. Therefore, the U.S. Supreme Court has ruled that simple penalty clauses are illegal unless the contractor, in the same contract, is given an opportunity to gain more monies by bringing the completion of the project in sooner. Thus, if the invitation to bid or the "General Conditions" of the specifications note that there is a requirement that the project be finished by a certain time, it may list a daily charge against the contract for all time after the specified completion date. However, if the contract documents include a simple penalty clause, they must also include an allowance *to the contractor* for each day he improves on the specified completion date.

There is, however, another method that is used to insure close adherence to the required completion date without offering the contractor extra money for completing the project early. This is the **"liquidated damage"** system wherein the owner states that, unless the project is completed by the specified date, he will lose X dollars per day. The contract documents will further state that the owner will expect to be repaid this daily amount by the contractor for every day between the specified date and the date when the owner actually may take beneficial occupancy of the new facility. A liquidated damage clause, factually backed up by loss figures, does not have to offer the contractor opportunity for extra payment for early completion. Parenthetically, however, we should all realize that the sooner a contractor or construction organization can bring a project to completion, the more profit that is available to the builder! Thus, regardless of contractual clauses, it is to the builder's best interests to complete any contract **as soon as possible!**

We have now discussed Lump-Sum, Cost-Plus Fixed-Fee, and "upset-sum" Cost-Plus Fixed-Fee contracts. We have also discussed contracts that penalize the construction organization for not completing on time and provide a bonus for completing early. We have shown where the owner may gain or lose in each type of contract. However, until we had listed and described each type of contract, we could not discuss all the favorable and unfavorable aspects of contract types until we had discussed **all** the contractual systems.

One might feel that the Lump-Sum contract is advantageous to the owner only. Nothing could be further from actual fact. To explain, we must discuss one of the biggest problems that confronts a builder with other types of contracts. Generally, most Cost-Plus contracts require that the owner or the owner's representative be requested to give permission for everything that the builder does. That is, the owner must agree to the use of special machinery, special allowances for better subcontractors, and to overtime. If the builder does not request prior permission, he may find that the owner will refuse to pay for certain expenditures.

For example, a Project Superintendent is often confronted with a decision on overtime. And, quite often, the decision to keep workers on the project for additional time may save twice the overtime premium wages expended by making the project ready for the next operation sooner. A Project Superintendent may decide that, if he keeps carpenters and concrete reinforcing installers working overtime on a Thursday, he will be able to make a concrete pour on Friday. This will allow the concrete to cure on Saturday and Sunday (i.e., nonworking days), which will, in turn, allow the forms to be stripped two days earlier. Under any of the Cost-Plus contracts, prior approval by the owner or his representative would be necessary because the superintendent might be spending the owner's money. And, quite often in these situations, the owner is reluctant to agree to an extra expenditure. Thus two days of curing time are lost at a time when it might be very important. This type of lethargy or indecision on the owner's part could cost the builder considerable money!

However, under a Lump-Sum contract, all construction expenditures come out of the builder's funds and, thus, the superintendent is free to make any decision that he feels will be advantageous to the project without waiting for an owner's representative to "second guess" him. Thus, if construction designs are completed, the Lump-Sum contract can be more advantageous to the builder because it gives him full control of his company's potentials. And, as the owner cannot lose on a Lump-Sum contract (when the design drawings are complete), this is a most equitable type of contract for all concerned.

There is one "if" in the contract-type decision that usually determines which type of contract the owner or contractor might desire. If the design

drawings are substantially completed at contract time, the Lump-Sum contract is usually preferred. If they are not, one of the Cost-Plus Fixed-Fee
contracts will be used.

3–8   Additions to Contracts

There have been a number of cases when the contract drawings were so
accurate and so complete that there was no necessity for additions and/or
changes. However, and unfortunately, this is rarely the case. Changes come
about for a number of reasons. The owner may find that he has additional
needs, the designer may have omitted a requirement, or conditions such as
extra rock in excavating operations or continuous inclement weather may
entitle the contractor to additional funds.

Regardless of the type of contract, there are provisions in each to allow
for adjustment of contract terms. Under usual circumstances the contractor
will advise the Architect that the unanticipated field conditions or the additions shown on the revised drawings constitute an "extra." He will list the
items involved and he will price each. If the owner or his Architect does not
agree that the extra charges are fair, there will be negotiations until an
agreement is reached. At that time the Architect will issue a formal **change
order** to the General Contractor. This change order will list the plan changes
or the field conditions with an agreed-upon price against each. This change
order form will be signed by the Architect and, then, by the owner. It will
then be sent to the General Contractor for countersigning (whereafter at
least one signed copy is returned to the Architect). If some of the extra work
is to be accomplished by a subcontractor, such as the excavation subcontractor or the foundation subcontractor, the General Contractor will issue
change orders to the subcontractors involved. These change orders will be
additions or deductions to the original contract. If the original contract is
subject to a time-completion clause, the change order will note whether
additional contract time is to be allowed.

3–9   Periodic Payments, Less a Percentage

Consider. If one General Contractor bid $10,100,000 on a project and
your contracting firm bid $10,500,000 for the same project, there must be
good reasons for the difference. For purposes of our discussion these reasons
are not important. However, if the owner signed a contract with the low
bidder and if, for any reason, the original contractor had to leave the project
after 50 percent of the work, the owner would find it hard to find another
contractor to finish the work for the remaining 50 percent of the contract
amount (i.e., $5,050,000). **Your company** made a bid of $10,500,000. Surely
your company will not wish to pick up another contractor's work (and
problems) for $200,000 less than its original bid.

To explain, your company's original bid of $10,500,000 was $400,000

higher than the bid of the contractor who won the contract. Therefore, even
if his costs were correct, at the 50 percent stage he would have accounted
for $5,050,000; if your company's costs and profit portion for the second half
of the project (if the bid was accurately figured) would be half of its
original bid of $10,500,000, or $5,250,000—a $200,000 higher figure!

For this reason most contracts stipulate that, after the Architect's approval, the owner will pay the contractor *periodic payments* (i.e., payments
from month to month) to the extent of the contractor's costs **LESS** 10 or 15
percent. This percentage reduction of payment is called "retainage." Thus, if
in the situation we are considering, the first contractor could not continue
his contractual obligations and complete the project, the owner and his
Architect could find another contractor who would. At this 50 percent point
the owner would have paid the following:

Amount due (or paid to) original contractor	
50% of material and labor, including fees	$5,050,000
Less retainage of 10% (per contract)	505,000
Actual amount paid to date	$4,545,000
Amount remaining in funds ($10,100,000 less $4,545,000 paid)	$5,555,000

Now, then, if **your** company had made an original bid for the entire
project in the amount of $10,500,000 and this was a carefully estimated
figure, your company would not be willing to complete another's contract
for 50 percent of the *original* contractor's price. However, as 50 percent of
your company's original bid is $5,250,000, and even if your company wishes
a bonus to "pick up somebody else's troubles," the owner has $305,000 over
one-half of your original figures to work with. This $305,000 "cushion" will,
probably, make it possible for the owner to complete his project within the
original budget. It is for this reason that most contracts stipulate at least 10
percent retainage until "substantial completion" of the project, when the
retainage percentage may be reduced. Of course, when the project is completed, the contractor may bill for 100 percent of the original contract, which
will include any retainage previously held.

We have now discussed the makeup of a construction organization and
its usual personnel. We have discussed the manner in which a new facility is
designed. We have discussed the different contracts under which a new
facility will be built. We shall now progress to the processes in the actual
construction of a new facility. In general, we shall discuss *first* the things
that a construction organization must do *first*. However, as we discuss the
actual construction of a facility from excavation, to foundations, to roofing,
we may pass certain aspects of construction and come back to them later
when, with more knowledge, we can discuss them more fully.

Chapter 4

Starting the Project

The construction business starts out as a business for young people. That is, many of the tasks to be accomplished in the actual construction operations must be accomplished by strong and vigorous people and, therefore, a young person finds that he can excel if he puts a strong body, along with his mind, into his effort. If he excels in the lower plateaus of the business, he finds he can find work in the management of a construction project and, thus, many young men have climbed the ladder of success to Assistant Superintendent, Superintendent, and Project Manager. The opportunity for a young person to proceed quickly to a position of great responsibility, where he may hold sway over a multimillion dollar project, is available in few other industries. The speed with which a person in construction climbs the ladder of success depends, in part, on vigor and ability. However, more importantly, success depends on **knowledge of the construction business.** Mere knowledge of engineering and construction, though most important, is not the final answer. In order that a person may handle a project to the gain of his employer and to his personal gain depends on precise and varied knowledge of the construction business.

Another reason that the construction business is attractive to young people is that it is continually challenging. No two construction projects confront the managers with the same problems. No two construction projects present the same goals. Each day brings different problems to be solved. Thus young people find the business fulfilling, and older persons find that the constant challenge helps to keep them feeling young.

But the key to **anyone's** success in this business is knowledge of the business. Our text has progressed through the foundations of the construction business and now it moves out into the field where the construction is actually accomplished. In the field the *actual construction* falls under the regulation of the Project Superintendent. Thus, even though certain portions of the preliminary preparations may have been accomplished by the main office, it is his responsibility to check that they *have been* accomplished and to take personal action to see that the remaining responsibilities are taken. Permits are, understandably, most important; and it is the responsibility of the Project Superintendent to satisfy himself that all required permits are obtained by those needing these permits.

Basically, two types of permits are required for a construction project, (1) those that must be obtained by the designers of the project, and (2) those that must be obtained by the builders of the project.

In some localities (usually principal cities) the designers and owners of the project may be required to make an initial presentation to the community's Planning Board, with preliminary sketches and "impact" studies to show the authorities that the new project fits in with the community's long-range planning. And, if the project falls close to or in a built-up neighborhood, its designers may have to come before a Fine Arts Commission or similar body. If the project falls short of the community's intent, the owner will have to seek a "variance" which is legal permission for a project to differ from a community's zoning or building ordinances in certain aspects as requested.

However, after preliminary approvals have been received from the community, the designers must make final design drawings for the architectural, mechanical, and electrical phases of the construction and present these to the Examiners of the community's Department of Buildings. These Examiners will check the structural and architectural design drawings to see that they fulfill the structural, architectural, and safety requirements of the local and state codes, and that the mechanical and electrical design drawings live up to local and state requirements. After the Architect and his mechanical/electrical consultants have demonstrated to the officials of the community that the project falls under all local and code requirements, the owner will receive permission for the project, and the Architect and the owner will proceed to take bids from construction organizations. When a construction organization is chosen and given a contract to produce the project in accordance with design drawings and specifications, the construction organization will proceed to seek and procure all the permits that are required of contractors and subcontractors. Now the work of the Project Superintendent begins.

Whereas the construction organization or its Construction Manager may have awarded a contract to a company that specializes in wrecking and has applied to the community's Department of Buildings for a building permit, it is usually the responsibility of the **Project Superintendent** to follow up to insure that the wrecking subcontractor gets a wrecking permit and that the initial building permit is given to his company. The name of the Project Superintendent and his home telephone number, along with his signature, must appear on some building permits.

Most communities give final permits for wrecking, the entire structure, and for plumbing and electrical construction at the beginning of the project. However, some communities issue preliminary permits for excavation and foundation work, but withhold permits for the erection of structural steel, superstructural concrete, masonry, and the like, until foundations and basement walls are completed to the building inspector's satisfaction. Thus, if the project is to proceed on schedule, it is the responsibility of the Construction Superintendent to see that permits required for structural work and permits required by mechanical and electrical subcontractors are sought and received as is necessary.

In addition, there are other permits for such items as crane placement, sewer connections, temporary electrical sub-stations, material storage, and fuel storage (to list a few) that must be sought and issued before each phase of the construction can proceed. ALL of these permits incur fees, which, according to the wording of the contract(s) will be borne by the General Contractor or the subcontractor. In addition, some communities require deposits or cash bonds for certain phases of construction and permits. An example might be the excavation under the paving of a street or material storage on a portion of the street. Before issuing a permit to a contractor, the community might require a deposit or bond to insure that, when construction work is completed, the pavement will be repaired as required and left in good condition. The range of requirements for permits varies greatly from area to area. However, the basic permits are required everywhere.

4–2   Insurance

States and all communities require several types of insurance to protect the workers, pedestrians who pass the project, neighbors of the project, and the community itself. Until the community receives certificates to show that the applicant is covered by sufficient insurance, it will not issue permits. Types of insurance that the state and the community most usually require are the following:

1.   Workmen's compensation insurance.
2.   Workmen's disability allowance insurance.
3.   Public liability insurance.
4.   Property damage insurance.

Workman's compensation insurance covers medical costs to treat workmen and, if a workman cannot return to work temporarily or finally, covers a weekly living allowance for the period he cannot work. The amount of the weekly payment is uniform and is set by law. However, the insurance premiums for the state-required coverage are dependent on (1) an insurance rate set in accordance with the accident rate of a certain trade, and (2) an insurance rate set in accordance with the accident rate of the **certain construction company.** Whereas a Project Superintendent and his field organization have little effective control over the statewide accident rate of a certain construction trade, he and his field organization can exert considerable influence on the **company's** accident experience rate. Because insurance costs are a significant portion of the costs of a construction project, the adverse experience rate of a construction company can affect the company to the point where it may lose a competitive bid to a company with a lower insurance rate.

Thus it is most important that a construction organization make every effort to insure the **safest project possible.** A construction project whose safety program is inadequately watched can very well put the company "out of the running" in competitive bidding for three years (which is the usual experience period used by construction-insurance companies). To help maintain as low an accident rate as possible, many construction companies hire safety engineers to supervise the safety aspects of their projects, or retain safety specialists to inspect their projects periodically and make written reports to the superintendent and to management.

The well-being and life of another human being should be sufficient incentive to prod any Project Superintendent and his forces to strive for a safe project. However, with construction costs ever rising, the superintendent and his field organization have another huge reason to strive, continually, for safety. Safety has now become a primary requirement of a good construction organization. Therefore, this text will **often remind** the reader of safety requirements and safe practice in construction.

Workmen's disability allowance insurance covers the worker when he is not on the project or for injuries not incurred by work in his regular occupation. Thus, although costs for such required insurance reflect upon the cost of a project, the Project Superintendent and his field organization cannot affect its cost by local actions. However, project records must indicate that such claims were not job incurred.

Public liability insurance covers workmen not in the employ of the contractor and other parties not involved in the construction. Because any claims made will be attributed to the insured, it is important that a construction organization make every effort to keep equipment, hoists, cranes, vehicles, and the environment in which they are used in the best possible condition. And, if an accident is caused to such a "third party," it is important that accurate records be kept in case of future suit. In such cases the insurance carrier should be notified immediately in case it wishes to make other records or seek additional witnesses or information.

Property damage insurance, as its name implies, covers a contractor from claims in regard to damage done by his workers or the equipment used by his company (such as powder-activated devices, company trucks, or blasting operations). In most cases the insurance carrier will wish complete reports and/or photographs. Where blasting or pile driving is scheduled, the insurance carrier will wish to make preliminary surveys of surrounding structures to take official note and make photographs of any cracks or deficiencies in these structures **before** blasting or pile driving.

The amount of coverage for the different types of insurance discussed so far is usually set forth by the State and the community. However, these government-required limits may not be high enough in the Architect's or owner's opinion to protect the owner from claims. For this reason it is usual practice for the contract specifications to list limits which the owner will accept, which are sometimes higher than statutory requirements.

Fire and extended coverage insurance, although not a requirement of the civil or permit-extending authorities, is a most necessary requirement of the owner and his Architect. Thus the specifications will list the amount of coverage that the owner will require. Whereas a claim for fire will, when paid, replace the loss due to fire, and a claim for extended coverage (i.e., damage from wind, rain, or other "Acts of God") may pay the contractor for particular losses, the payments **will not** repay the contractor for other losses, such as time when project personnel could be working on a new project. Thus it is important that the Project Superintendent and his field organization bend every effort to protect the project from fire and inspect all closures, lashings, towers, and the like, to see that winds, rain, and other "Acts of God" do not cause damage. Therefore, as this text proceeds, the reader will often be reminded that safety and precautions for safety are most important.

4–3   Existing-Conditions Surveys

There are a number of "existing conditions" that the contractor may wish to survey and make a matter of record before he proceeds on a project. As a matter of fact, although he has been supplied with **boring charts** supplied by the designers to guide all contractors in bidding the excavation, a contractor may wish to take further borings or, at least, dig test pits before he makes his initial bid. In addition, he may note that the condition of structures adjacent to the project may need special protection. Thus his bid may be affected by prebid site inspections.

However, after a contractor accepts a contract, he will most certainly wish to make surveys and take photographs of adjacent structures that may or may not be the source of future claims. In this regard, as we noted under the discussion on insurance, he may wish to have the help and guidance of his insurance carrier. If his scheduled work includes blasting or pile driving, his insurance carrier will insist on existing-condition surveys and photographs. In addition, if his scheduled work includes pile driving, he will want to set up a surveyor's level and mark level lines on **the actual cross-hair** line of the level on adjacent structures, so that future surveys made while pile driving is in progress will indicate, most accurately, if any settlement has been caused to occur on these peripheral structures. Quite often these preliminary and final existing-condition surveys prove that there have been no changes and, therefore that no damage has been sustained. However, without the preliminary surveys (and perhaps photographs) the contractor and his insurance carrier would not be protected from untrue claims.

In addition to protection from future damage claims, existing-condition surveys may forewarn the contractor or his Project Superintendent of conditions that may change the method in which the project will be handled. For example, if preliminary investigations indicate that storage areas are tight,

the contractor may decide that it would help to pay for redesign and extra
construction costs to make the first floor framing stronger so that delivery
trucks could back onto that slab with their loads. Or the contractor might, at
least, decide to rent adjacent areas before actual construction started and
his company's presence on the site tended to raise rental prices. These
maneuvers will be expanded on in later chapters.

4–4 Initial-Layout Survey

Before excavation for a project and the pouring of footings can commence, there must be an initial layout of the project. Most construction
companies have field engineers who are accomplished surveyors, so they are
well able to work from **monuments,**[1] **curb cuts,**[2] or other location points
shown on the Architect's site drawing. Whether the *initial* layout is done by
the construction company's field engineers or by a licensed surveyor is a
question that the Project Manager or the Project Superintendent will decide.
The decision will be made on the basis of whether the **legal** accuracy of the
building lines is important. Most surely, a good surveyor on the Field Engineering staff of the construction company is capable of such a layout and
the continuing line-and-grade checks that will fall to his duty throughout
the life of the project. Therefore, if the building line falls well within the
owner's property so that a slight misinterpretation of a surveyor's base line is
not critical, the entire layout, both initial and future, will be made by the
construction company's Field Engineers. However, if the face of the building ("F/B") is set by the Architect to coincide with the property line, it
would probably be wiser to retain a licensed surveyor who has considerable
professional experience in the city and in the area of the project to set down
the basic building lines (which coincide with city property lines or the
property lines of neighbors), and to make initial curb cuts so that the
construction company's engineer can always reproduce them. In most cases
this licensed surveyor will be the surveyor who made the original survey for
the owner. The notes he has shown on his original survey drawing and the
additional notes on his calculation sheets will make it possible for him to do
the most accurate survey. However, more importantly, the construction
company is relieved of considerable legal burden if the original plot lines
are set by an independent, licensed surveyor.

[1] A monument, as used in construction work, is a concrete or stone shaft inserted
into the earth so that its bottom end is below frost line or at least deep enough (where
frost is no problem) so that it will not be disturbed. If the monument is stone, a mark is
cut into its top with a cold chisel. If the monument is cast-in-place concrete, a nail is set
into its top while the concrete is soft. These marks denote property corners or survey lines,
as required. See Figure 4-2.
[2] A curb cut is a most usual mark of a surveyor or Field Engineer. Basically, it
is a line or cross cut (with a sharp cold chisel) into a curb adjacent to or across the road
from a building. As the concrete or stone curb is fairly permanent, the mark will endure.
If the mark is to be used for line (only) it will look like ⋔ . If the mark is to serve as a
reference for line *and* a distance from the mark, it will look like ✛ .

Consider the building shown in Figure 4–1. Our client has purchased a

piece of property on the northwest corner of Main Boulevard and Second Street. The city's zoning regulations in this area require that any building be set back from the curb line a distance of at least 20 feet. Our site plan shows that the sidewalks on Main Boulevard are 20 ft wide, and it shows a line on the Second Street side that indicates this **building line**.[3] Note also that the Architect has placed the western building line ("B/L") so that it coincides with the property line of the grocery store land. This line and the Main Boulevard building line are important. If the building, when finally erected, overhangs the city's property (i.e., the city building line), this will constitute a violation. If the building overhangs the grocery store property, this fact may not be discovered until the grocery store property is sold and the new owner elects to build on the entire site. Actually, the unhappy fact that the building had passed either of the property lines would probably be discovered at the time of the resurvey required by most Building Departments. Regardless of when it was discovered, it would be a very costly situation for the builder to alleviate. Considering that a licensed city surveyor who was well aware of the location of property and street lines in the area could stake out building lines that would not be disputed and that his charges will be fair, the project should have original staking by a licensed surveyor. While this surveyor is working on the site, it should not be too costly for the builder to retain him to put curb cuts on the main building lines (including the center line of the building) so that these cuts will always be available when stakes, batterboards, or other marks are damaged. We have shown the "curb cuts" on Main Boulevard at points A and A', B and B', and C and C'. We have shown curb cuts on Second Avenue at points D and D' and E and E'. At the time that these lines are being laid out, the surveyor should drive stakes near the northerly property line on lines AA', BB', and CC'. He should put a stake somewhere at the westerly end of the rear building line (i.e., the EE' line), and he should put an offset line on the grocery store's easterly facade so that a monument or batterboard can be placed on these lines by the builder's Field Engineer. Finally, the licensed surveyor should present the builder with a drawing of the area that shows the intended location of the new facility. On this drawing he will show all the curb cuts and stakes that he has set out, and he will reference (i.e., show the distance from the new building line) curb cuts A, B, and C. We have shown this referencing as a distance "d." This referencing is important. It will provide the builder's Field Engineer with points he can easily and quickly measure to new construction from time to time and provide him with an easy way to "watch" his main building line as construction progresses.

[3] The Building Line or "F/B" is the line on which the face(s) of the building will fall. When this line falls on the legal Building Line of the municipality or on the property line between the owner of your building and an adjacent owner, this property-building line should be staked out and marked by a licensed city surveyor.

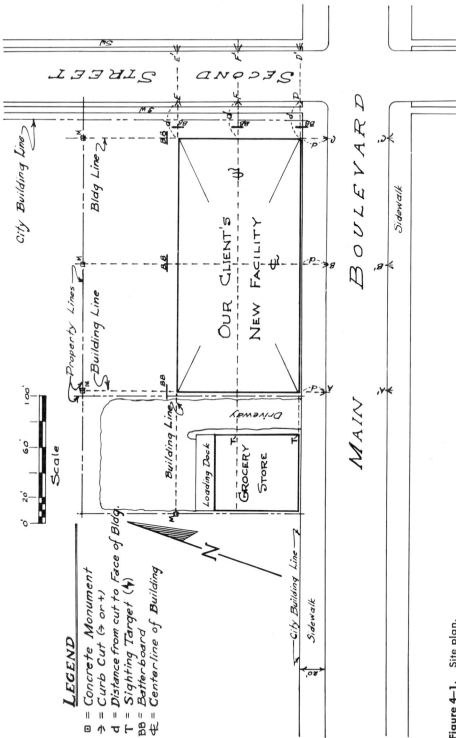

Figure 4-1. Site plan.

LEGEND

▣ = Concrete Monument
→ = Curb Cut (→ or +)
d = Distance from cut to Face of Bldg.
T = Sighting Target (↟)
BB = Batterboard
⌷ = Centerline of Building

Scale

0'   20'   60'   100'

Regardless of whether the initial layout of building lines was made by the builder's Field Engineer or by a licensed surveyor, there will have to be additional reference marks, batterboards, and monuments so that the Field Engineer(s) and foremen on the project have more easily used reference lines. A "monument" when used in surveying is either a stone or a concrete shaft, usually 6 by 6 in. or 8 by 8 in., and at least 36 in. long so that it will not be heaved by frost in the winter, which has a mark in its top to show surveying lines (see Fig. 4–2). If the monument is stone or precast concrete, the engineer or surveyor must take considerable pains to insure that the monument is carefully set in well-tamped backfill; and then he will drill a hole into the top of the monument on the survey line. A poured-in-place monument is set by placing two stakes on the survey line approximately 18 in. on each side of the desired location for a monument (i.e., 36 in. apart) with nails or tacks on the survey line. After these two stakes are set, a laborer can take a posthole digger and excavate an approximately 6-in.-diameter hole, 36 in. deep, about halfway between the stakes. He can then fill the hole with concrete, place a reinforcing rod or two in the soft concrete, and top the monument with an 8- by 8-in. box or a piece of a No. 10 can as a form. As the top of the monument concrete sets up, a string can be run between the stakes' nails, and (with the use of a plumb bob) a nail can be set into the top of a monument. If the Field Engineering crew has a helper who is careful and is handy, the crew may set stakes for a number of monuments at one time, leaving the helper to excavate and pour the monuments and to place the nails in the monuments. The engineering crew

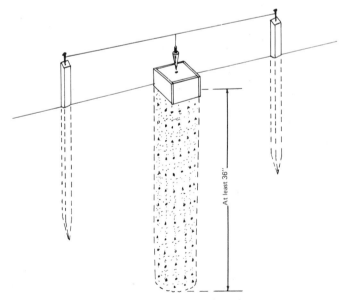

Figure 4–2. Concrete monument (poured in place).

should find time to check the nails in the monuments (from the original lines) before the concrete becomes too hard to adjust the nails. This type of monument is not difficult to set, and if the survey line is important (and there is space), it is worth setting one. However, if there is a building on the survey line and behind the site, a semipermanent and adequate reference mark may be placed onto that building. The marks are semipermanent because, after permission has been granted by the owner of the other building, the marks must be set so that they may be easily removed after they have ceased to be useful.

A **batterboard** is a semipermanent type of marker for building lines. See Figure 4–3. Although we say a batterboard is "semipermanent," it is more correct to say that a batterboard is much more permanent than a stake (which could be knocked out of line with one blow) because it has at least two heavy members driven deeper into the ground and it is well braced. Thus a blow to one of the two by fours will be resisted by both two by fours. The cross brace is an integral and most important part of the batterboard. Without it the batterboard is merely two heavy stakes. And, because the batterboard is steady and can be higher than a stake, nails for center line (₵) of column or face of building (◄) can be driven into the top rail and carpenters can hang their lines between batterboards. If the batterboards

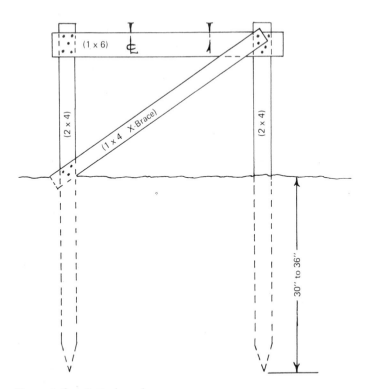

Figure 4–3. Batterboard.

are made wide enough, they can be used to mark column center lines and, after the columns are erected, the nails can be moved 24 in. to one side so that the engineer and the carpenters will have a 2-ft AXIS line.

4–6   Placement of Other Reference Marks

The project shown in Figure 4–1 is, of course, hypothetical. However, all projects are different, and this project situation is sketched to show a number of the problems the Field Engineer will have to face. Basically, it is an **urban** situation. Therefore, there is not much room to play with. We have suggested that curb cuts be placed in locations shown as A to E'. The sidewalk on the Second Street side of the project is narrow. Thus we can have batterboards between this sidewalk and the top of the excavation. There is, approximately, 70 ft between the new building and the rear property line. This space will probably be allocated by the Project Superintendent as shanty and sawmill space. Column center lines or **axis lines**[4] can be placed on the (top) fronts of these structures, and batterboards can be set for places where there is no structure. Regardless, there should be at least one concrete monument at the rear of the lot. We have shown three. However, as the project proceeds, two may be out of sight. The Field Engineer or the Project Superintendent should seek permission from the owner of the grocery store property to set a monument on the back line of the building (i.e., the E line) and to place a batterboard on the same line adjacent to the driveway.

In addition, the 20-ft sidewalk on the north side of Main Boulevard will have to be removed, or a 15-ft portion removed as excavation proceeds. Batterboards placed along the front portion of the project (i.e., the Main Boulevard side) will not endure as long as those on the back and on the Second Street side. They will have to be replaced from time to time. However, because we have curb cuts on both sides of Main Boulevard, this will not be a difficult chore.

---

[4] Before a building starts to rise out of the foundations a batterboard, curb cut, or monument, can be set on the center of column ($\mathcal{C}$) line. These can be used to set the foundations for each column and the center lines to set anchor bolts for steel columns or for centering forms for concrete columns. However, after the columns have been set the Field Engineer cannot use these lines alone because the columns will obscure the sight line. Therefore, as the building rises, the Field Engineer will set a line onto the building, which is *parallel* to column center line and a certain distance from the column line. This is termed an "Axis Line." Usually, these lines are 24 in. off the column center line or halfway between two center lines. In addition, for poured-in-place concrete structures, an axis line is usually set 24 or 36 in. behind the outside face of concrete (F/C) so that the carpenters have an accurate line with which to set their forms. If the Field Engineer is experienced and thinking ahead, he will have batterboards built wide enough so that, after the center line of column cannot be used, he can measure at least 24 in. to one side on two opposite batterboards and set up axis-line nails.

We have discussed the legal reason why it is often wise to use the services of a licensed surveyor for initial layout of the building. Our reason has been that we should not take chances that the building would be outside (in this case) the Main Boulevard building line or outside the property line on the west. If we are sure that the surveyor has accurately placed these **design** lines we are still not "out of the woods." There is a possibility that our workman may *build* the structure slightly out of plumb. Therefore, if the Project Superintendent is wise, he will instruct the Field Engineer to move the nails of his batterboard (those for the building lines on the west and south) at least ½ in. in (i.e., toward the center of the building). In so large a building, ½ in. will never be missed. However, the ½-in. allowance will give a more normal working tolerance for the carpenters and bricklayers. And, even if these two facades are to be prefabricated material, the tolerance should be used. In this same regard, the Field Engineer will be wise to check the structure and the facade on all four sides at every lift (or rise in story height) and to place center lines on each and every column as they are finally set.

We have discussed the duties of the Project Superintendent and the Field Engineer in the initial stages of the project. In Chapter 5 we shall continue with a discussion of the many primary decisions the Project Manager and/or the Project Superintendent will have to make in the very early stages. Then, in Chapter 6, we shall drop back to a discussion of the surveying that has already started and will continue for the life of the project.

Chapter 5

Temporary Facilities and Their Location

5–1 Field Offices

The size and location of a construction organization's Field Office depends on the size of the project and what functions are to be covered in the field office. If the company's field personnel are to be limited to a Project Superintendent, a Field Engineer, and a timekeeper, the office may be a 8- by 16-ft shanty or office trailer. It is assumed, then, that all correspondence to the Architect and subcontractors, along with other correspondence, will come from the main office. However, if this work is to be done in the field, additional space will be required for the Project Manager, at least one accountant, and one or two secretaries. Regardless of its function, a field office should be within sight of the project if it is not on the site. And, if more than four people are to work out of this office, flush toilets should be provided.

For purposes of our discussion, we shall consider the same project on the corner of Main Boulevard and Second Street. (See Fig. 5–1.) So that we can discuss the *major* possibilities, we shall stipulate that all correspondence and other contractual work be done in this office. However, under usual situations, the work of the Project Manager and his staff will not be done in the field until after the project has been running several months. In the meantime, he will handle the paperwork from the main office.

If there is an empty store or office on the north side of Main Boulevard, over the grocery store, or on Second Street, this space will have been discovered by the Project Manager during the initial investigation of the project, and will have been rented as soon as the construction organization was notified that it was the successful bidder. If such space, along with toilet facilities, is not available, the company will have to build suitable facilities or set up rented office trailers as soon as project progress allows. Chemical toilets are available for indoor and outdoor use, but flush toilets along with at least one lavatory are much more satisfactory, especially if one or more female office workers are included in the staffing. Therefore, the office must be located where plumbing can run to the city's sewer system. Although the office should be durable and comfortable, it should not be built to last for

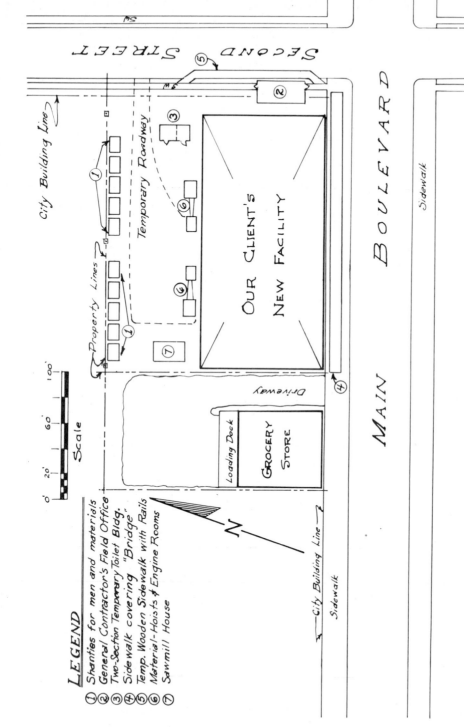

Figure 5–1. Location of facilities.

35

the length of the project unless there is sufficient room to leave it there. Usually, when building space is crowded, the field office is moved into the building as soon as the building is ready for temporary or final partitions. Figure 5–1 shows our temporary field office on the corner of Second Street covering a portion of the sidewalk.

In most cities a builder will be allowed to replace a portion of a sidewalk with a planked sidewalk. In some cities, regulations allow a construction organization to seek a permit to close a sidewalk on one side of a narrow street. Also, most cities will allow a sidewalk on such a wide street as Main Boulevard in front of a long facade such as ours to be replaced by a planked, covered, and electrically lighted temporary sidewalk. This sidewalk may be in the first 6 ft of the street in many cases. When the foundations and basement walls have been poured and when the front basement wall has been backfilled, this temporary sidewalk will be moved inside the curb if the permit so stipulates.

5–2   Workmen's Shanties

Our company will have to provide shanties for those workmen on the company payroll and will have to provide space for the workmen on the payrolls of subcontractors. Although trailers are available for this purpose, shanties constructed of 4-ft panels are most usually used. In projects where room is limited, temporary shanties are more desirable because they use less space per usable square foot and because they can be dismantled and moved into the building when progress allows. Shanties must be provided by the General Contractor for his *own* men and for materials that require security or weather protection. The subcontractors' men and materials will also require shanties. In most cases these subcontractors supply their own shanties. Sometimes, however, the General Contractor provides the shanties and charges the "subs" rental. In either case, shanties will be required for laborers, carpenters, concrete workers, plumbers, steamfitters, and electricians. If brick construction is to start in the very early part of the project, an additional shanty must be provided for these men. However, on buildings with more than one level, this shanty and shanties for additional men can wait until the **second** concrete floor (or arch) is poured. On a project where exterior space is at a premium, such indoor locations will be considered for these men who come later in the project. For our purposes, we have shown shanties and a temporary sawmill (which will be used to prefabricate concrete forms) along the northerly property line of the project. Note that space is allowed between each to lessen fire risk.

5–3   Temporary Sanitary Facilities

As soon as men report to the project, temporary toilets must be available. The law does not allow the first few men to depend on toilets in an adjacent gas station or restaurant. Thus the use of chemical toilets is most prevalent.

These are available, on a rental basis, from companies that specialize in supplying chemical toilets and providing weekly maintenance for them. However, for the convenience of all concerned, these should be replaced by flush toilets as soon as possible.

5–4   Provision of Temporary Services

Temporary services will be required from the time the project begins. Initially, temporary electrical power will be needed. Thus, as soon as the project starts, the Project Superintendent should order the electrical subcontractor to request a temporary tap from the power company and to run this power to the shanties, field office, and other locations as they are required. In most cases the specifications and/or the subcontract with the electrician state that the cost of providing this service is borne by the electrical subcontractor. The cost of the electrical energy is borne by the General Contractor.

Whereas provisions in the plumber's contract do not usually require that he provide temporary plumbing, they usually state that, after installation, the plumbing subcontractor shall maintain all temporary plumbing lines. The municipality may allow temporary water for fire protection and temporary constructional needs to be tapped (by the plumber) off one side of a fire hydrant. However, the plumber must *eventually* excavate the street for **permanent** water and sewer lines. Therefore, the superintendent should order him to excavate for and to install these lines **as soon as progress on the project allows.** The municipality will require that water used on the project be metered in any case. Therefore, if the water supply taps shown on the design drawings are installed, there should be no additional cost to the project. Also, the supply of water for fire protection and temporary use will not be limited. The sewer connections shown on the design drawings can usually be brought into the project soon after main excavation for the building has been completed in that area. The plumbing subcontractor should be ordered to excavate for and to install permanent sanitary and storm sewer lines as soon as possible. If the (design) storm sewers and required manholes are installed while the building excavation is in progress, there will be places to run lines from pumps that will remove water in the excavation. If the main sanitary sewer line is installed as soon as possible, it may be possible to set up a temporary flush toilet and lavatory facility for the men as soon as the superstructure allows.

5–5   Temporary Roads and Storage Facilities

All construction projects require coordination of delivery roads, the project, and storage areas. The superintendent had best decide on these locations immediately so that a load of material is not inadvertently placed in front of a delivery route. On all projects there must be access to truck earth from the basement excavation, and there must be access to all sides of the project where the contractor desires to pour concrete directly from the truck. For our project (see Fig. 5–1) we have shown the excavation route off Second Street. However, if there were a considerable grade differential on Main Boulevard so that the excavation were much shallower on the south side than on the Second Street side, the excavation ramp would come out of this side (within the confines of our property) onto Main Boulevard. Any ramps within the final foundation area will be removed by backhoe (Fig. 5–2) or bucket crane.

Storage facilities will be set wherever they do not impede construction. As soon as the structure rises, the superintendent will allocate space inside the structure. In this regard, if a structure is to be multistory and if delivery and storage facilities are difficult to locate the construction organization may consider asking the structural designer to increase the size (or strength) of street-level girders, beams, and their supporting columns for a certain portion of that level so that fully loaded trucks may back onto the floor. In many high-rise buildings, superstructure concrete is mixed right in the building in basement-located concrete mixers. On others, transit-mixed concrete rises from the concrete trucks via hoists. If street space is limited, the cost of minor redesign and the cost of "beefing-up" a portion of the first lift will be highly overshadowed by the saving on the delivery of the structural material and supplies right into the building. Note, however, that this is only economically feasible on a multistory or high-rise building.

Figure 5–2. Backhoe. (*Courtesy of Poclain*)

5–6 Location of Cranes and Hoists

In the last two decades great changes have evolved in mobile cranes. Previously, the use of cranes (as opposed to hoists) was restricted to areas where there was plenty of "swinging room" for the crane and its boom. Now however, the use of "tower-cranes" is becoming more prevalent because

these cranes may be set much closer to the building. Nevertheless, even on a project using one or more cranes at the start of a project, the use of **hod hoists**[1] will eventually be necessary. Whereas we have shown a hod hoist on the back of the building and a second location (shown in Fig. 5–1) for hoist location after the ramp is removed and the basement walls are back-filled, these locations depend on many variables, and the Project Super-intendent must decide on final locations. In some cases hod hoists are placed inside the building. However, the wise superintendent will **never** consider placing a hod hoist in an elevator shaft. The installation of perma-nent elevators takes **many** months, and if a hod hoist is placed in one of the shafts it will delay installation of that elevator (and usually an adjacent elevator) until it is removed. In such cases the elevators are rarely finished in time for use. The superintendent would be wiser to leave a shaft space in a certain portion of the floor slab at each level and patch the area later. This will give the project an interior hoist that can be replaced by a temporary *elevator* cab as soon as the first permanent elevator is substantially com-pleted. Parenthetically we should note that whereas temporary changes in a project's design **must be approved** by the Structural Engineer of Record, knowledge of stresses, as discussed in Chapter 17 will prepare the reader for choosing areas for such service.

A typical hod hoist and tower crane are shown in Figure 5–3 (a) and (b).

(a)   (b)

Figure 5–3.   (a) Material hoist. *(Courtesy of Patent Scaffolding Co., a division of Harsco Corp.)* (b) Tower crane. *(Koehring Crane and Excavation Group)*

---

[1] Years ago bricks and mortar were carried in hods on laborers' shoulders. Now all material is lifted with temporary elevators or hoists. Because of past systems, they are usually termed hod hoists.

The schedule for the construction of any project will allow for a certain amount of time (or number of days) when construction will not progress due to inclement weather or the effects of inclement weather. "Into **everyone's** life a little rain must fall." This period of time (or percentage of the entire schedule) will be based on the annual *average* rainfall of the area as compiled by the nearest branch of the Department of Commerce. We discussed this rainfall, previously, in Chapter 3. However, even though a company plans for a certain number of "shut-down" days, it must plan for many more days when inclement weather *will not* be allowed to shut down the project or, rather, when the company will take measures to handle the rain or cold and continue operations at the same time.

Two of the first things that come onto a project from a contractor's supply yard are pumps and tarpaulins. As soon as excavation is started, there are (usually) pockets or footing excavations that will be filled with rainwater or surface drainage. These must be pumped out. If the plumbing contractor has already installed the exterior portion of the storm-drainage system, the discharge hoses of these pumps could go to the storm-water manholes. If not, the water must be pumped over the curb or to areas that are satisfactory. During the initial stages of a project, tarpaulins will be needed to protect certain stored materials and newly poured concrete. Later in the project, tarpaulins may be used to surround areas where concrete or masonry operations need protection from rain or cold.

In addition to making provisions for *wet* weather, the schedule and estimate for construction will consider temporary heat in climates where outdoor temperatures may be expected to go below 32° Fahrenheit, or for times when a contractor will require 50 to 65°F temperatures to "cure" concrete or masonry mortar. In such cases, temporary heat will be provided by temporary heaters, blowers, or even steam coils. The heat will be kept within the desired areas with tarpaulins or polyethylene barriers. When temporary heating procedures will be extensive, there may be provision in the contracts of the H.V.A.C. (heating, ventilating, and air conditioning) subcontractor and the electrical subcontractor to supply this heat. However, on most projects this heat is supplied by the trades that require it. Of course, if there is water in plumbing lines, these lines must be protected by "frostproofing" (i.e., pipes are wrapped in insulation); in addition, electric *"tracing cables"* are sometimes wrapped around pipes before the insulation is placed over them to be sure the pipes are never as cold as 32°F.

5–8   Safety Measures

One might wonder why, in the early chapters of a text that covers *preparation* for construction operations, we should consider safety. Year by year, the safety program and the accident quota become more important to a construction project and to the entire construction industry. There is **no time** that is too early to plan for safety. This text will mention safety and safety programs **often** and a good construction man will make safety an integral part of his work. Thus, in the early planning for a project, a construction organization should decide if the project is large enough to require the services of a full-time safety man or whether periodic visits (and reports thereon) of a safety engineer will be effective. Regardless, the Project Superintendent should start an effective safety program as soon as he starts the project. Hard hats should be supplied to each workman, and signs that note that the wearing of hard hats is mandatory should be posted in the areas of workmen's shanties and in all other areas where workers or construction representatives may pass.

Second, the construction organization and/or its insurance carrier will check the local medical, hospital, and emergency facilities, and will post the names of doctors, hospitals, and ambulance services along with telephone numbers in the field office. Next, first aid supplies and equipment will be placed on the project, and personnel who are knowledgeable about first aid will be designated.

Finally, plans to **keep the project safe** will be formulated. Regular safety meetings, held periodically from the start of the project, will be scheduled. Attendees at these meetings will include a representative of each trade (or subcontractor), and the meetings will be chaired by the Project Superintendent or the Safety Engineer. At each meeting the Chairman will remind the attendees about routine safety precautions, such as the wearing of hard hats, replacement of temporary barriers, and the use of safety equipment. Representatives of the trades may mention unsafe conditions that the superintendent or the Safety Engineer may not have observed. All these items will be carefully considered, and minutes of each meeting should be prepared and copies should be given to each man attending and to his employer. If minutes are prepared and issued, the construction company will have records to show the safety requirements of the project, and to show that all workers (through their designated representatives) have been advised of these requirements. In the unfortunate case of accidents and the lawsuits that may follow, minutes of regular and effective safety meetings will do much to protect the contractor.

Years ago some construction superintendents often kept safety precautions to a bare minimum to save their time and their employer's money. Their safety was a "chance" operation. Some superintendents kept better safety precautions for humanitarian reasons. However, even in those days, a rise in insurance rates (eventually) penalized contractors with high accident rates, and "credits" cut the rates of contractors whose accident experience over a specified period (usually three years) was lower than the norm. Thus, even in the early days, it was financially profitable to keep a safer project.

Now, however, safety protection has become increasingly required. There have always been State safety requirements and safety legislation.

The effectiveness of a State's safety program depended on the efficiency of its inspectors and the manner in which they followed up on the violations noted. Because the efficiency of programs varied, the Federal Government set up the Occupational Safety and Health Administration (**OSHA**) as a subdivision of the Department of Labor. This agency set up rules for the construction industry and for **all** industries, and appointed inspectors to check actual conditions and levy fines where necessary. At the time that **The Contract Work Hours and Safety Standards Act** created OSHA, it stipulated that any State which could prove that it had an effective safety program and wished to run its own safety program could do so after application was made by the State and permission given by the Federal Government. Some States are acting on this provision.

Regardless, SAFETY—at the start of a project—throughout the project—to the end of a project—is a "MUST."

Chapter 6

Surveying on the Project

Any student of construction practices and *all* Project Superintendents should have a working knowledge of surveying and the manner in which it is accomplished. Thus the reader is urged to go to a technical library and borrow a text on plane surveying. The word "plane" refers to a totally flat surface. Thus plane surveying is surveying that considers a portion of the earth as a flat surface as opposed to the more complicated system of surveying needed for work on larger areas where the curvature of the earth would have to be considered. The reader will find that such a text contains detailed diagrams of surveying instruments and gives instruction on the theory of angles, balancing of computations for traverses, computation of acreage, computation of excavations, and the accuracy tolerances to which a surveyor must work.

Surveying and any of the subjects we discuss in this text require hundreds of pages and many lessons for complete instruction. This text touches on the highlights of each subject and the special needs for construction. Thus in this chapter we must assume that the reader has a working knowledge of surveying basics. We shall point out the areas where special care is required and shall set forth basic constructional surveying practices. The actual study of basic surveying must be conveyed by surveying texts and surveying teachers. Thus, even though a construction man has a working knowledge of surveying instruments and their use, he would do well to take a text on surveying practice and brush up on his skills.

6–1   Care of Surveying Instruments

The transit and the level are most precise instruments. (See Fig. 6–1.) We shall assume that they were in good adjustment when they were delivered to the project. But they must be handled with care and protected from any abuse that will increase the need for adjustment or might actually damage them. These instruments are enclosed and shipped in cases that closely surround them and that have cushioned guards to grasp and protect each part. Thus, when the instruments are not being used, these cases are

(a)

(b)

Figure 6–1. (a) Surveyor's level. (b) Surveyor's transit. (*Courtesy of Kueffel & Esser* Co.)

the safest places for them. Certain construction companys build racks so that instruments can be left on their tripods and stored for ready use. This is a labor- and time-saving device. However, it is a system that we recommend **be avoided.** Instruments, even if covered by the plastic rain covers available, are subject to dust situations that can cause bearing damage and lens damage. And if an instrument should fall from the rack, great damage can occur.

The minor extra time that will be expended by taking the instrument from its case when it is required and replacing it in its case after each use is well worth the safety it affords.

The instrument should be continuously attended from the time it is set on its tripod until it is replaced in its case. The instrument man should stand near it, continuously, even though he is writing in his field book or making measurements in the vicinity. He should always be prepared to keep his instrument from falling. In addition, whereas the covers provided by instrument manufacturer to protect the instrument from rain should be used for this purpose, they should *not* be used in windy conditions. At such times these covers act like sails so that the wind will push the instrument over. If the instrument man is ceasing operations for lunch, he may set the tripod legs wide if he can watch the instrument during his lunch. Otherwise, he should replace it in its case until he returns to work.

6–2   Adjustments to a Surveying Instrument

There are a number of adjustments to a surveying instrument that can be made in the field. Included are correction of level bubbles, adjustment of cross hairs, adjustment of verniers, disassembly of plates for cleaning and lubrication, and several others. If one was using an instrument far from a major city, he might **have** to do all this work. However, if the instrument is being used where it can receive adjustments in an instrument shop and where a "loaner" instrument may be rented while yours is being adjusted, the *only* adjustment recommended in the field is that on the level bubbles, which is quick and easy. Other adjustments are time consuming and are more quickly and accurately accomplished in the controlled environment of a nearby instrument shop. Adjustment to level bubbles is easily accomplished by (1) leveling the instrument, (2) rotating the instrument 180°, (3) adjusting the bubbles with the adjusting pin (provided in the instrument case) to 50 percent of the noticed error, and (4) repeating steps 1, 2, and 3 until no further adjustment is necessary. This adjustment which is fast and easy may have to be made from time to time on both levels and transits. However, although adjustments to cross hairs and verniers *can* be made in the field by an experienced instrument man, they are not recommended.

6–3   Accuracy of Surveys

Surveying is a process that may or may not be accurate, depending on the person making the survey. In most cases it is the methods used by the surveyor or Field Engineer, rather than the accuracy of the instrument, that determines the accuracy of the survey, because a knowledgeable instrumentman can take special care to allow for the possibility that his instrument needs adjustment. In fact, a good instrumentman will take these special precautions at all times by assuming that his instrument might have fallen out of adjustment.

Sometimes it is necessary to make a quick, one-shot location so that construction may proceed immediately. However, as soon as time allows, the Field Engineer will make further checks or do additional surveying to prove that this location was accurate or to make a slight correction if he finds an error. He may have needed the location quickly at the first instance, but he will repeat his survey before the item he located is "poured in concrete."

In using a transit to set angles and a tape to measure distances to these angles, the only way to be sure that all the angles and distances are correct is to run the complete **traverse**. A traverse is a system of angles and distances that returns to the point of origin. A simple traverse is shown in Figure 6–2.

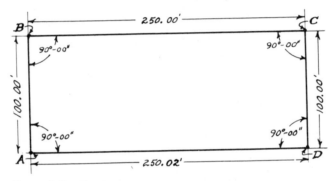

Figure 6–2. Traverse.

If the Field Engineer set his transit on point A and sighted onto his other property-line mark in line with point D, turned 90° and set point B at a distance 100.00 ft from point A, he would have set his first angle (D-A-B) and would have two points on his final traverse. He would then continue by moving his transit to point B, backsighting on A, turning 90°, and setting point C on this 90° line 250.00 ft away. Continuing, he would move his transit to point C, backsight on point B, and set point D 100.00 ft from point C. He now has stakes at points A, B, C, and D, with tacks set accurately on the line and by the distances measured. But he has not closed his traverse. If he sets up his transit on point D, turns a 90° angle (i.e., 90°-00'-00", or exactly 90°) and the instrument's cross hairs fall exactly onto the tack at stake A, he knows that his first three angles and distances are probably reasonably accurate. If he now measures the line AD and finds the distance to be 100.00 ft, he **knows** that all four angles and all four distances have been accurately measured. However, if the closing distance, A-D, measures 100.02 ft (an error of 0.02 ft or, approximately, ¼ in.), he will know that there is an error in one of the four angles (or a smaller error in each of the four), or that distance B-C is incorrect. The latter possibility is probably the cause, because angular errors (unless counteracting) would not allow the cross hairs to hit the tack at point A. If the cross hairs **do** hit the final point (A), then there must be balancing errors because, by plane-geometric theory, all interior angles of a geometric figure (in this case a four-sided traverse) must add up to $n - 2$ (where n is the number of interior angles) times 180°. In our traverse this would be $(4 - 2) \times 180° = 360°$. If there were more than four sides to the traverse, it might be more difficult to spot where the error occurred, and one might have to rerun the traverse. However, when we discuss the manipulation of the transit, we shall discuss the methods of assuring that angles are turned accurately and proved at each point (or corner).

A similar closure system (which one might term a vertical traverse) should be used in the setting of all important bench marks (reference grade marks). To explain this system we must, first, consider the workings of a surveyor's level, and we must assume that this instrument and its user are exactly accurate. If the instrumentman sets up his level and adjusts the bubbles so that the telescope is level and then sights a rod on a point of known elevation (such as a city bench mark) of elevation 10.960 ft and reads a "backsight" (+ shot) of 4.24 ft, his height of instrument ("H.I.") is 10.960 + 4.24 ft or 15.20 ft. If a "foresight" (− shot) is taken on an unknown elevation and the rod reading is 6.28 ft, then the elevation of this new point is that of the "H.I." minus 6.28. Therefore, the elevation of the new point would be 15.28 − 6.28 = 8.92 ft. Such a system is fine for finding a quick grade. However, if the new grade is going to be used as a bench mark for the project, one must close the traverse and come back to the original starting point, as we did on the horizontal traverse.

Consider the notes from a Field Engineer's field notebook shown in Figure 6–3.

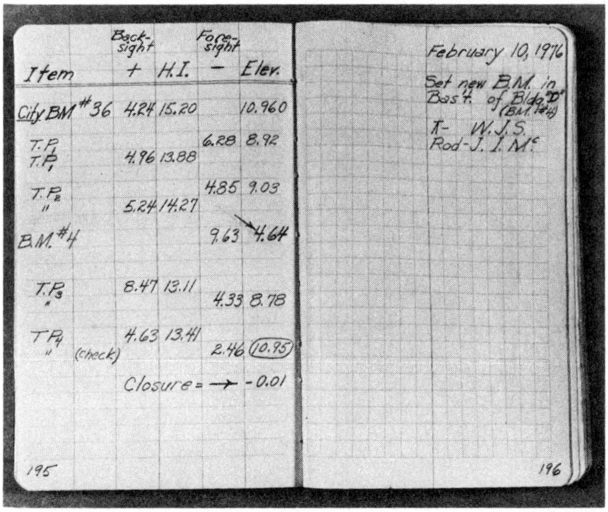

Figure 6–3. Field Engineer's leveling notes.

Our engineer wished to set a bench mark in the basement of building

"D." He set up his instrument and took a backsight ($+$) on a rod set on
City Bench Mark No. 36 (which has an actual elevation of 10.960 ft), and
he read 4.24 ft. This gave him an H.I. elevation of $10.960 + 4.24 = 15.20$ ft.
He then took a foresight ($-$) on a temporary point (perhaps a bolt on a fire
hydrant) and read 6.28 ft. This point (called a **turning point**) had an
elevation of $15.20 - 6.28 = 8.92$ ft. This turning point he noted as TP_1. He
then moved his instrument to another point (toward his goal) and, when it
was leveled, read a backsight ($+$) of 4.96 ft. Therefore, his H.I. was the
elevation of TP_1 (8.92) plus 4.96, or 13.88 ft. He then sighted forward to a
second turning point (perhaps a marked point on a curb) and read a
foresight ($-$) of 4.85 ft. As the H.I. was 13.88 ft, the elevation of TP_2 was
$13.88 - 4.85 = 9.03$ ft. He was then close to the new bench mark he wished
to establish. He lifted his instrument and moved approximately halfway
between TP_2 and the bench mark (BM) he wished to establish (Project
BM 4). After leveling his instrument, he read a backsight of 5.24 ft, which
gave him an H.I. of 14.27 ft. He then took a reading on the rod, which had
been moved to BM 4, and read 9.63 ft. Thus the elevation of BM 4 was
$14.27 - 9.63 = 4.64$ ft.

The Field Engineer could have stopped here and considered his surveying accurate and, therefore, that the elevation of BM 4 was actually 4.64 ft.
However, as he was setting a bench mark that would control a portion of
the project's elevation, he continued his surveying back to the point of
origin, the original City BM 36. His next setup had a backsight reading of
8.47 ft, which gave his setup H.I. as 13.11 ft. He then took a foresight on a
new turning point (TP_3) and the reading of 4.33 ft. established the elevation
of TP_3 as $13.11 - 4.33 = 8.78$ ft. One more move was necessary. He moved
his level approximately halfway between TP_3 and his starting point (City
BM 36) and read a backsight of 4.63 ft. As this was a backsight ($+$), his
H.I. was established at $8.78 + 4.63 = 13.41$ ft. From this setup he took a
foresight on City BM 36 and read 2.46 ft. This set the elevation of City BM
36 as 10.95 ft. However, note that the actual elevation of City BM 36 is
10.960. In the entire vertical traverse there was an aggregate error of 0.01 ft
(or slightly more than ⅛ in.). One should note that the original city bench
mark was accurate to three decimal places (i.e., $\frac{1}{1000}$ ft). Had our Field
Engineer been carrying this traverse for a longer distance or if he needed his
new bench mark to be accurate to $\frac{1}{1000}$ ft, he could have used a target with
a vernier (see your surveying text), and would have read his backsights and
foresights to $\frac{1}{1000}$ ft. However, on most construction projects $\frac{1}{100}$ ft is
accurate enough, more quickly read, and yields a quick, accurate, bench
mark. Also, if we had taken this sample calculation to one more decimal, it
might have led to confusion for the reader. As it is, we have shown the
foresight and H.I. for a level setup on a different line than the (next)
backsight and TP elevation. Most engineers would have the foresight, H.I.,
backsight, and elevation of the TP on the same line. However, for a first
instruction in level use, our system is more clear.

One should note that, in balancing errors on closed circuits of elevations or in closed transit-tape survey traverses, the Field Engineer must choose whether he wishes to average the errors or to rerun the survey. The amount of error is his guide in these cases. In the example shown in Figure 6–3, the engineer closed his circuit within 0.01 ft. At the worst, his new BM 4 is within $\frac{1}{100}$ ft (approximately ⅛ in.) and, more likely (if the total error is an aggregate of minor errors at four setups), it is within 0.005 ft of the correct elevation.

6–4   Accuracy of the Surveyor's Level

A surveyor's level can be one of the most mis-used instruments because it is an instrument that most construction men can quickly understand. Thus a carpenter foreman may borrow the Field Engineer's level to set a few grades on a wall form. If the level is not accurate or if he uses it poorly, the grades he sets may be inaccurate. However, with the same inaccurate instrument, a knowledgeable Field Engineer could set grades that were much more accurate.

Consider Figure 6–4. To illustrate a point we have shown the sight lines (whose angle from level are exaggerated) going from both directions. The student should understand that it is the intention of the diagram to indicate sight lines when the telescope is turned from the first shot (backsight) to the next shot (foresight).

Notice that, if there is an inaccuracy in the sight line (in this case caused by incorrect cross hair), the sight line will not be level and, thus, the reading of a backsight will not be accurate. If the backsight distance was 200 ft and the sight line was tilting downward x inches, then the H.I. would seem to be x inches low. However, if the foresight was the *same* distance (i.e., 200 ft), the error of y inches would be exactly the same as the backsight error (x inches) and would, thus, compensate the measurements. The foresight reading will be exactly x inches too small. And, because the H.I. is x inches too low, the (balancing) incorrect foresight will compensate the error and give a calculated elevation for the foresight that is correct. However, if a foresight was considerably longer than the backsight, there would be an error z that is greater than x or y. Were this the case, the elevation of the foresight grade would be calculated low by a dimension of $z - x$.

The point to be remembered is that, if the level is placed so that foresights are (approximately) equal to backsights, errors caused by instrument inaccuracy will *compensate* themselves. Thus, if the carpenter foreman sets his level at a good distance from the wall form that needs grades, and if he were extending grades from a known grade (4-ft mark) on one end of the forms every 10 ft to the other end of the forms, his grade line would be fairly accurate. However, if he sets his instrument closer to the forms, there will be a dip (or a rise) in the center portion of the line, because the instrument is closer to the center of the forms than to the ends of the forms.

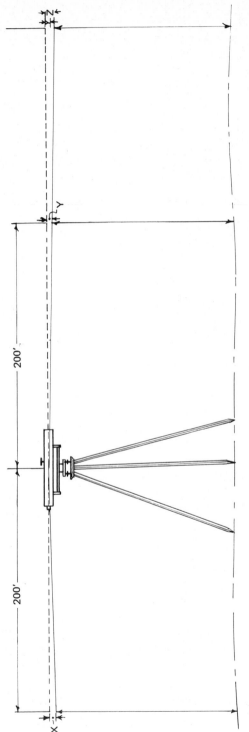

200'

200'

X

Y

Z

Figure 6-4.

The experienced Field Engineer will *assume* that his level might be inaccurate or, at least, use it so that any errors which may be caused by a recent mishap will compensate themselves. In review, a good engineer can get accurate results even if his instrument is inaccurate. Conversely, a less experienced engineer can get poor results with a most accurate instrument if he is not careful.

6–5   Accuracy of the Surveyor's Transit

The same general system of errors that we considered in the discussion of the level are possible with the transit. However, in the case of the transit, we shall find that the errors are in the angle of the line rather than the level of the line. We should note that, if a transit is used to transfer grades (i.e., the transit is used as a level), the situation will be the *same* as with the level. However, because a transit's telescope is shorter than a level's telescope, and because there is a possibility of vertical change in the transit's telescope, it is considered poor practice to use a transit to project accurate levels or grades.

In regard to errors that may be found in transits, consider Figure 6–5, in which we are looking down on the top of the transit. The sight lines are drawn to indicate a situation where a backsight is taken with a transit set onto a stake (tack), which is on a particular line to a distant point on the same line. Then, *without turning any angle,* the instrumentman merely vertically rotates the telescope (on its horizontal axis) from a backsight to a foresight. This is called *plunging* the transit and is a most usual operation. If there is an inaccuracy in the cross hairs, there will be an angular error equal which causes a linear error d in one direction. However, if after marking a point for the first shot (which is incorrect by a distance d), the transit telescope is left in its (upside down) foresight position and rotated 180°, reset to the backsight point, and plunged forward for a new foresight, there will be an error equal to a d distance in the *opposite* direction.

Thus, if the transit is plunged twice with a 180° reset turn, the Field Engineer will end up with two points that are a distance apart equal to $2d$. If he then measures the distance and puts a tack (or mark) halfway between the two points, the tack (or mark) will be exactly correct; if the engineer wishes to extend the line a few feet farther, he may set his cross hair on the new mark and extend the line to a new point.

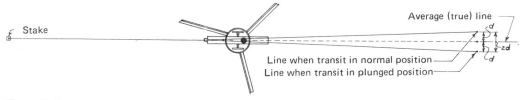

Figure 6–5.

A similar situation is experienced when an engineer wishes to raise a point on a wall (or batterboard) to a point higher on that same wall, which will require that the telescope of the transit be raised considerably in angle. If the leveling bubble on the front of the transit's plate is not quite accurate, the transit's telescope would not raise on a line that is plumb. However, rotating the telescope 180°, plunging the telescope so that the bottom of the wall can be sighted, and lifting the line in a second (plunged) operation, will give an error in the *opposite* direction. The engineer can then split the difference between the two points and know that the new (third) mark is absolutely accurate. He can then sight his transit on the new (halfway) mark and lower it 1 or 2 in. so that he has two points for a sighting target. Even if a knowledgeable Field Engineer does not find any noticeable error in raising such a sight point, he will always use a two-operation plunge-system sighting to raise sight lines vertically. By so doing he always has a check on himself and his instrument.

Finally, we should consider a system of additive turning of angles, which is primarily used in construction to be sure that the angles are correct regardless of telescope error and one that is used by engineers and surveyors to read an angle more accurately. Most of the transits used in construction work read to degrees, minutes, and the closest half-minute (i.e., nearest 30 seconds). However, if the Field Engineer needed to read an angle closer to 30 seconds he would repeat the angle for two, four, or six times (with additive readings) and divide the *total* angle by the number of times he had "run up" the angle. To more clearly understand this process, the reader should consider a situation where the transit is set over point *B* and the user wishes to measure the angle between two other points, *A* and *C*, from the setup point. That is, he wishes to measure the angle *A-B-C*. If he wishes to measure the angle to a fair degree of accuracy, he will release the set screws for both the bottom plate and the top plate, set the vernier to 0°, and lock the top plate's set screw. He will then move the telescope until the cross hair is on point *A* and lock the bottom plate's set screw. Thereafter, he will loosen the top plate's set screw and turn the telescope to point *C*, tighten the set screw, and read the angle after turning the slow-motion screw for the upper plate.

Let us assume that the instrumentman read 78°–10′–30″, plus a few seconds. If this angle reading was close enough for him or if he did not wish to check this angle, he would write 78°-10′-30″ in his book and continue with his survey. However, he would more probably write this angle in his book and then "turn up" the angle four times. To do this, he would (without disturbing the reading of the upper plate) loosen the bottom plate, set the telescope back on point *A*, lock the bottom plate, loosen the top plate, and move the telescope to *B*. After a slow-motion adjustment on the upper plate (so that the cross hair is accurately on *B*), the reading on the vernier should read somewhat more than twice the first reading, probably about 156°-21′-00″ plus. However, after *two more* repetitions of this process (four turn ups), the total angle will read 312°-42′-30″ (exact on the seconds). He will then divide this **"run-up"** angle by four and find that a more accurate reading of angle *A-B-C* is 78°—10′—37½″.

If the instrumentman was positive that his transit was absolutely accurate, he would run up the four turns of the angle with the telescope in the normal position. However, if he is taking the trouble to turn up the angle two, or four times, he will invert the telescope for half of the settings (in our example, twice). In this way any error of the cross-hair or transit mechanisms will be compensated in the manner we previously discussed.

Now, if a Field Engineer has to set points by 90° angles that should be most accurate, he will set the point on a 90° line and then run up the angle four times (with two in plunged position). If the total run-up angle does not read exactly 360°—00′—00″, he will move the point slightly and run up the angle again. Angles for a building's layout (see Fig. 6–2) should be run up four times.

6–6   Basic Conclusions on Accuracy

As we noted before, this text cannot teach surveying. However, in discussing the use of the surveyor's level and transit, we have endeavored to show that there is a way to make accurate settings even if the instrument is inaccurate. In fact, a Field Engineer should use his instrument so that if it is slightly out of adjustment since last used, the settings will be accurate. If after checking an instrument he finds that it needs adjustment, he should have the instrument adjusted. Nevertheless, he should continue to take precautions after the instrument is returned, even though he is now sure that it is accurate. These precautions take very little more time and they provide a check on the instrument *and* the instrumentman.

In the building business there is a saying that "a poor workman blames his own tools." This should not be allowed. In the Field Engineering portion of the business, poor layout *cannot* be tolerated. A good engineer can achieve good results if he knows his business!

Chapter 7

Scheduling for a Project

A schedule is something we have all used from our earliest days. A personal schedule for many men might list a 6 A.M. rising, shaving at 6:15, breakfast at 6:30, start to work at 7:00, and arrive at work at 7:45 A.M. Because this is a personal schedule and a fairly usual schedule, it is seldom documented or set forth in charts.

However, when a schedule must cover many complicated activities, such as a construction project embraces, the schedule must be carefully planned, precisely set forth (or documented), and systematically followed. The schedule for a construction project must consider the ordering (and delivery) of materials and the construction processes wherein these materials are erected or installed.

In the scheduling and purchasing of construction materials the builder must deal with normal-supply items and long-lead supply items. **Normal-supply** items are those which are usually stocked by local suppliers and can be obtained quickly. Typical would be concrete, concrete block, lumber, and roofing materials. **Long-lead** items are those which will take several months to obtain after an order for the item has been received. The length (or time), or the "lead," is determined by such things as the time to process shop drawings and to get the order into the manufacturing schedule of the fabricator.

Therefore, in scheduling a project the builder must determine how many weeks will be used for excavation, installation of footings and foundation walls, erection of structural steel, masonry, and everything else that is to be a part of the completed project, including the paving and landscaping. In doing this, he must *also* consider how many months will be needed between the time when a material or item is ordered and when it is delivered to the project. The most obvious long-lead item is structural steel. The steel fabricating shop must prepare shop drawings, which in addition to listing the lineal weight (per foot) and shape type of each steel member will give the exact length and the connection details. After preparation, these drawings must be checked and approved by the Design Engineer and then entered onto the production schedule of the fabricating shop. If the project is using a steel fabricator who does not manufacture the steel shapes, there must be a time allowance for ordering of the shapes from the rolling mill. However, even if the steel is to be fabricated and erected by one of the large companies that still smelt iron *and* roll structural steel shapes, the company's fabrication department must schedule its fabrication to the company's shape-rolling schedule.

Because the structural steel portion of a project comes directly after the foundation stage of a building, the order for structural steel should be placed several months before excavation is contemplated. Then, if the fabricating of structural steel, the excavation operations, and the foundation operations fall within the activity lengths[1] of the project's schedule, the first loads of structural steel will be arriving at the job as soon as there are foundations for this steel. Thus the ordering of structural steel and its delivery scheduling constitute a long-lead problem that is easy to comprehend. Because the builder will not wish to keep his superintendent, engineers, and foremen idle for any period between the completion of foundations and erection of steel, he will order his steel early and will push field production so that foundations are finishing just as the structural steel arrives at the job site. Then, just as soon as a portion of steel is erected and finish-bolted, he can move his crews onto decking and the subsequent construction operations.

Other long-lead items that the builder must schedule are such items as special brick, hollow metal doors and door bucks, sash, architectural metal items, precast concrete, and many others. These are items necessary for the structural portion of the project. However, of equal scheduling importance are the long-lead items of the plumbing, H.V.A.C. (heating, ventilating, and air conditioning), electrical, and elevator subcontractors. Where the equipment to be installed by these subcontractors has to be installed (or at least moved into position) before a portion of the building may be erected, this equipment must be scheduled to arrive in time.

7–1 The Bar-Chart System of Scheduling

The bar-chart system of charting a construction schedule has been in use for many decades in construction and is, therefore, easily comprehended by most construction men. It shows, visually, the proposed starting and completion dates for each operation. Because the sequence of operations is shown on a parallel type of graph, an error in original thinking is easier to find. In most cases a project's bar chart does not show the sequence of ordering on long-lead items. However, as a superintendent is given the plans, specifications, and production schedule for his project, he is told if the

---

[1] An "activity" is any single operation in the construction. Three examples would be shop drawings for steel, poured concrete foundations, or even set entrance sleeves. The **activity length** is the period of time for any particular activity.

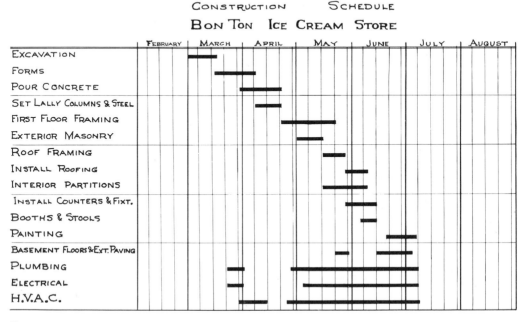

structural steel is on schedule and when he must be ready for the erection of the first stage or derrick of the structural steel. As he takes over his duties in building the project, he will be checking with the company's Purchasing Agent to be sure that orders have been placed in time to insure that face brick will be ready when he needs it and that door bucks will be delivered to the project several weeks before partitions are to be erected. Week by week he will be checking with his Purchasing Agent and subcontractors to insure that materials are scheduled to arrive when his project will need them.

Figure 7–1 shows an example of a bar-chart schedule. For purposes of a simplified example we have chosen a small, basement plus one-story building that has several long-lead items. Because the basement walls are poured concrete, the builder must allow some time for preparation of reinforcing-steel shop drawings and their approval. There are Lally columns and floor-supporting steel girders that must be detailed, approved, and fabricated before erection. And the construction portion of the contract includes counters and soda-fountain accessories, which must be ordered in time for installation. The mechanical/electrical portions of such a small building will not have too many complicated items. However, most of the equipment to be installed by these subcontractors is subject to the submission of brochures and "cuts" and their approval before ordering.

Because this is a very small project, the preparation of reinforcing-steel shop drawings should not take too long; thus delivery of the shaped reinforcing steel should be on the project about 15 days after it is ordered. If

Figure 7–1. Bar-chart schedule.

Lally columns and the minor amount of structural steel come from local

sources, no more than 35 days should be needed from time of order to time
of delivery. Thus, if the shop drawings for these two items are in progress
when excavation starts or slightly before, the reinforcing steel should be on
the project as soon as footings are to be poured and cellar walls are formed.
Thereafter, the Project Superintendent and his main-office people must confer from time to time to be sure that all the other materials and trades fall
into the original schedule. This would be the bar-chart schedule and system
for a small building such as the Bon Ton Ice Cream Store. However, the bar
chart and system for a multimillion dollar project would be similar. As
previously noted, it is a simple system that is easily comprehended by most
construction men, from the Project Managers and Superintendents right
down to the foremen who must back them up.

7–2   The Critical-Path System

In the first portion of this chapter we reminded the reader that we have
all been following one sort of schedule or another since we demanded our
first feeding of milk. We discussed a man's daily schedule from the time he
waked at 6 A.M. until he arrived at work at 7:45 A.M. As listed, this is all
the information that one would need for this man's personal bar-chart
schedule. However, consider that when the man started shaving certain
other actions had to be completed previously. When the man went into the
bathroom to start shaving, there had to be (previously purchased) shaving
cream there. And the day before the wife went into the kitchen to cook the
breakfast, she had to buy some or all the breakfast items.

If the householder who does the purchasing can remember to have
enough food in the house each morning, then a person's morning schedule
can be plotted with a bar chart. However, if one must preschedule the buying of orange juice, coffee, and eggs, then a schedule system like the Critical-
Path Method (CPM) must be used. The scheduling of a project follows
similar needs.

Because the CPM system takes so many *more* operations into consideration, a printed example in this text will take more space. For this reason,
along with the fundamental requirement of a simple example, we chose a
simplified project like the Bon Ton Ice Cream Store to discuss with **both**
systems (see Fig. 7–2).

The line diagram for the Bon Ton's CPM looks complicated. However,
basically it is not. From point 1 which is the start of shop drawings for all
trades (including the mechanical/electrical trades) and, in this case, the
start of excavation, to point 25, which is the conclusion of the project, the
activities are scheduled so that operations of work (activities) that **must**
precede others are so scheduled, giving sufficient time, before later activities
are scheduled.

For example, before forms can be set (activity 2–3), excavation (activity
1–2) must be completed. And before basement (cellar) walls can be poured

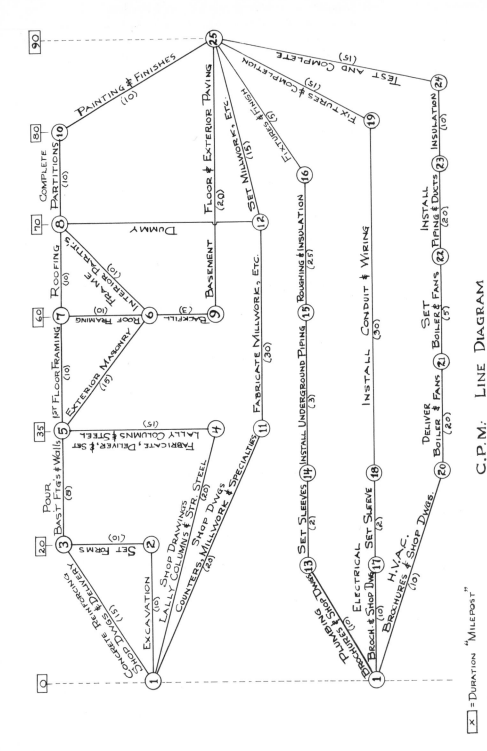

C.P.M. LINE DIAGRAM

BON TON ICE CREAM STORE

x = DURATION "MILEPOST"

Figure 7–2. CPW line-diagram.

(activity 3–5), concrete reinforcing shop drawings must be approved, steel

fabricated (activity 1–3) and form setting (activity 2–3) completed. In this
regard, note that there is a dotted line between the number 1 in the top
portion of the diagram and the number 1 in the bottom portion of the
diagram. This indicates that it is the *same* point (1). Were we to start the
mechanical/electrical from the same number 1 in the diagram, it would be
confusing. However, the reader must understand that all activities starting
from a 1 start at the same time.

If we continue on the upper portion of the CPM, we note that (as shown
on the bar chart) the first floor framing follows the substructure work.
However, before this framing (activity 5–7) can be set, shop drawings for
Lally columns and steel (activity 1–4) and steel fabrication and delivery
(activity 4–5) must be completed. After the exterior masonry (activity 5–6)
and the first floor framing (activity 5–7) are completed, roof framing (activity 6–7) can commence and framing of interior partitions (activity 6–8) can
commence. While these activities are in progress, the construction contractor's items, such as counters and soda-fountain equipment, should have
been through the shop-drawing stages (activity 1–11) and coming through
fabrication (activity 11–12).

Similarly, and at the same time, mechanical/electrical subcontractors
should have been processing shop drawings, cuts, and brochures, and placing sleeves and inserts into the construction contractor's forms so that, as the
need arose, preparation for their portion of the work could be continued.

You will note on the line diagram that there are maximum time allowances for each activity. Thus, if excavation took the maximum of 10 (working) days and the setting of forms took the maximum of 10 days (a total of
20 days), concrete reinforcing would be ready and installation of reinforcing
could be set as soon as sufficient forms (activity 2–3) were set. The listed
maximum activity times should be laid out with discretion and knowledgeably. Therefore, activities *should* be completed within their allotted times.
However, when conditions change, the schedule must change. And, in all
cases, schedules must be rearranged to give final results or as close to final
(schedule) results as is possible.

For example, this line diagram was laid out so that, while construction
went through points 1–4–5–6–7–8–10–25 (in this case the **critical path**), the
other activities continued and (hopefully) were accomplished and completed within the same time frame. The reader will note that the total of
certain activities (such as the millwork, plumbing, H.V.A.C., and electrical)
adds up to less than the 90 working days required for the activities that fall
on the "critical path." This allows later starts or, perhaps, more activity days
for certain work. These extra days are termed *"float"* and are shown in
Figure 7–3, the computer run for this diagram. However, if plumbing rough-
ing (activity 15–16) were delayed for any reason, an early-start "float"
might be exceeded and the furring of pipe chases (in interior partitions)
would also be delayed. Also, counters and millwork (activity 12–25) might
have to be expedited if the project were to finish within the 90 working days

PROJECT CONTROL SYSTEM—BON TON ICE CREAM STORE

Activity	Description	Duration	Start Early	Start Late	Finish Early	Finish Late	Total Float
1-2	Excavation	10	March 1	March 8	March 14	March 21	5
1-3	Concrete Reinforcing—S.D. & Del.	15	March 1	March 8	March 21	March 28	5
1-4	Lally Cols & Steel—Shop Dwgs.	20	March 1	March 1	March 28	March 28	0
1-11	Counters, Millwork, Etc., Shop Dwgs.	20	March 1	March 22	March 28	April 19	15
1-13	Plumbing—Brochures & Shop Dwgs.	10	March 1	March 29	March 14	April 12	20
1-17	Electrical—Brochures & Shop Dwgs.	10	March 1	March 22	March 14	April 5	15
1-20	H.V.A.C.—Brochures & Shop Dwgs.	10	March 1	March 15	March 14	March 29	10
2-3	Set Basement Forms	10	March 15	March 22	March 29	April 5	5
3-5	Pour Footings and Basement Walls	8	March 29	April 9	April 10	April 19	7
4-5	Lally Cols & Str. Steel-Fab. & Inst.	15	March 29	March 29	April 19	April 19	0
5-7	Install 1st Floor Framing	10	April 19	May 10	May 3	May 24	15
5-6	Erect Exterior Masonry	15	April 22	April 22	May 10	May 10	0
6-7	Install Roof Framing	10	May 10	May 10	May 24	May 24	0
7-8	Install Roofing	10	May 24	May 24	June 7	June 7	0
6-8	Frame-Out Interior Partitions	10	May 10	May 24	May 24	June 7	10
6-9	Backfill	3	May 13	June 5	May 16	June 10	17
8-10	Complete Interior Partitions	10	June 7	June 7	June 21	June 21	0
9-25	Bas't Floor & Exterior Paving	20	May 20	June 10	June 17	July 8	15
8-12	Dummy	0	June 21	June 21	June 21	June 21	0
10-25	Painting & Finishing	10	June 21	June 21	July 5	July 5	0
11-12	Fabricate Millwork, Counters, Etc.	30	March 29	May 3	May 10	June 14	25
12-25	Set Counters, Millwork, Etc.	15	May 10	June 14	June 3	July 5	25
13-14	Plumber—Set Sleeves Bas't Walls	2	March 22	March 29	March 25	April 2	5
14-15	Plumber—Underground Piping	3	April 22	May 14	April 25	May 17	16
14-16	Plumber—Roughing & Insulation	25	May 10	May 24	June 14	June 28	10
16-25	Plumber—Install Fixtures & Fin.	5	June 24	June 28	June 28	July 5	5
17-18	Electrician—Set Entrance Sleeve	2	March 22	March 29	March 25	April 2	5
18-19	Electrician—Conduit & Wiring	30	May 3	May 3	June 14	June 14	0
19-25	Electrician—Fixtures & Complete	15	June 14	June 14	July 5	July 5	0
20-21	H.V.A.C.—Deliver Boiler & Fans	20	March 29	April 12	April 26	May 10	10
21-22	H.V.A.C.—Set Boiler & Fans	5	April 26	May 10	May 3	May 17	10
22-23	H.V.A.C.—Install Piping & Ducts	20	May 3	May 17	June 3	June 14	10
23-24	H.V.A.C.—Insulation	10	June 3	June 17	June 17	July 1	10
24-25	H.V.A.C.—Test & Complete	15	June 3	June 17	June 24	July 8	10

Figure 7–3. Information contained on CPM computer run sheets.

(18 weeks). Thus the new critical path might fall along the 14–15–16–25 line or the 10–12–25 line. In the latter case a device (termed *dummy*) is used to connect diagrams so that they can be fed into the computers. This line from one diagram point to another has no time designation, and, as it is not an actual activity, is called a dummy connection. On this diagram it is activity 8–12.

Because the critical-path system is divided into so many parts and shows all the details of the project, it has the advantage that it can be rearranged when project conditions change with the intent that the project will still be completed within the same allotted time. And because the activity times are indicated by the number of working days, this system can be fed into a computer which, *when accurately updated*, will note the problems and indicate the new critical path.

If the project started on March 1, the earliest date that reinforcing could be finished would be March 21. However, it will take a total of 35 days for the shop drawings for Lally columns and structural steel (20) and fabrication of these items (15). And it will take a total of 20 days for excavation (10) and setting of forms (10). Therefore, there is an allowable "float" of 5 days for the reinforcing (20 − 15) and, because there *could* be a float between the *total* activity times 1–2–3–5 and 1–4–5 of 7 days [(20 + 15) − (10 + 10 + 8)], these 7 days plus the 5-day float between 1–2–3 and 1–3 (12 days total) would originally be allotted to activity 1–3, and the remaining 7 days would (originally) be allotted to activity 3–5. Or, to take another alternative, portions of certain other activities could be started earlier so that, when the rest of the project has "caught up," they will also finish in sequence.

For example, on a recent project, because of a shortage of sheet-metal workers in the area, the contractor set up his schedule with sheet-metal installation as his "critical path." Thus this pending threat to the schedule was shown in the computer runs. In the same schedule the concrete slab-on-grade (actually a slab on piles) was shown with considerable extra time or float. The Project Superintendent noted this and elected to push the concrete slab ahead of schedule. This smooth concrete surface allowed the sheet-metal workers to use rolling scaffolds instead of the pole scaffolding that had been envisioned in the original schedule. Thus the sheet-metal subcontractor could use less men to do the same work, and this subcontractor finished his work in less days than were originally anticipated. The moving of the critical path to a route through the concrete floor saved many days on the project.

Thus the CPM scheduling system is a great tool. But like many tools, it is relatively useless unless it is kept "sharp." One keeps a CPM sharp or accurate with regular updating. This is accomplished by *regular* meetings of the superintendent (and perhaps some of his assistants) and the computer specialist. At these regular meetings the superintendent and his assistants review each current activity with the computer specialist. After each of these meetings the computer specialist corrects the input for the computer and makes a new computer runout. The latest runout tells all concerned the current status of the project and which pathway is critical, so that the project will be completed on time or as close as possible to the original finish date. However, the CPM is not as good a tool as it could be if the original thinking on the schedule was not correct or if it is not regularly checked and updated. In most cases regular updating is the key. If the best CPM is not regularly checked and updated it tends to become useless.

7–3   Choosing a Scheduling System

Anybody can easily visualize the bar-chart system. Regardless of its size (as determined by the size or complexity of a project), it remains forthright. On the other hand, a CPM network diagram for a large project can be complex to the point that (because of the many print-out sheets involved), a mistake in *original thinking* can evade many.

Thus, if the project is not too complex, a bar-chart schedule whose scale is large enough to make each working week at least 1 in. long can be a most useful tool. If space is allowed under each bar for *planned* activity length so that the superintendent can draw a line for *actual* activity length, the schedule becomes a daily updated tool that is posted on the office wall for all to see and learn from. Also, because the schedule is simply presented, many an old-timer who knows his business can pick up a mistake in original thinking made by a man with less construction background. True, the bar-chart system burdens the superintendent with the chore of assuring himself, from time to time, that material suppliers are meeting schedules and that his Purchasing Agent has covered all purchases.

If the project is very large or very complex, the CPM is the answer. However, if the CPM is to be a good scheduling device, it must be well made. First, the CPM starts (roughly) as a series of bar charts. Then each activity bar is carefully scrutinized for optimum time starts and durations. Initial diagrams are made, critical paths compared, and *then* the final diagram is made.

Another advantage of the CPM is adaptability to computerization. Payments to contractors and subcontractors are usually based upon price breakdowns submitted after a contractor has won a contract. These price or cost breakdowns can be set into a computer as functions of the CPM's accomplishment readings. The output will be a cost breakdown to be compared with contractors' requests for progress payments.

7–4   Recapitulation on Schedule Systems

Basically, *no* schedule is worth a "tinker's damn" unless it is heeded. Once formulated and accepted, it must be followed. As soon as people in charge of the project start to take the schedule loosely, they start to lose hold of their project. Therefore, pick the best scheduling system and then follow it well! And, if the Project Superintendent does not do more than "just follow" the schedule readings, he is not doing enough. There must be regular meetings at which the superintendent (or Project Manager) tells the other contractors the status of the project and where certain trades are behind. At these meetings some contractors may advise where expediting other's work will help them.

It's not enough to have a good schedule. One must follow the progress of the project and, when a portion of the project falls behind, take action to bring the project back onto schedule. Lost days on construction contracts cost **everybody** money.

Chapter 8

Excavation

Under the original General Contractor system, excavation was one of the trades that was accomplished by the General Contractor with his own men and equipment. However, when excavation machinery became more sophisticated and as hauling trucks became larger, this was one of the first subdivisions that went "subcontractor." However, although much of the responsibility is shifted onto a subcontractor, there are two aspects to remember:

1. Even though the subcontractor has most of the responsibilities for the *excavation* of the project, including project safety and the shoring of streets and adjacent structures, the Project Superintendent and his company will be held to be negligent and partially responsible if there is damage to other persons, adjacent properties, or the equipment of others.

2. After the excavation subcontractor has completed a portion or the whole of his contract, that portion completed becomes the responsibility of the Project Superintendent unless otherwise stated in the subcontractual documents.

Regardless of the stage of construction while excavation is in progress, the Project Superintendent and his company have most definite responsibilities to the owner and to the public for safety on the project.

We have considered protecting the owner and the public. Now let's consider protecting the excavation itself. Basically, there are two systems of excavation to be considered: (1) the main excavation for the building or structure, and (2) the excavations required for storm sewage and sanitary sewage. The excavation for the sanitary sewage may be scheduled for a time when it can be installed when required by the plumbing contractor and when it will cause less problems to the total construction effort. Usually, this is after the footings, peripheral basement walls, and the floor above the basement level are poured.

However, as soon as somebody has excavated a hole in the ground, it will become filled with rainwater unless there is a place to pump that water.

"Would you believe—the storm sewers shown on the contract documents?"
Sure, these storm sewers have to be installed *someday*. Excepting for some
urban projects, most projects have an underground storm-sewer system and
it should be used. Let's excavate for it at the same time that the basement
excavation is being done and get the storm drainage piping laid and the
manholes erected as soon as possible. That way, as soon as we have a large
puddle in the basement excavation, we have a place to pump the water.
Good thinking? Sure, but few contractors think this deeply. Be one of the
few. Think dry!

8–1 · Types of Excavation

Although there are many types of excavation and excavation materials,
for all practical aspects the materials to be excavated may be subdivided
into loose sand, loam and clay with small rocks, hard clay with or without
rocks, soft rock (loose sandstone, serpentine, and rock that may be excavated without blasting), and hard rock. Thus there are two aspects for the
excavation contractor and the Project Superintendent to consider: (1) shoring to keep any loose material from caving into the hole, and (2) shoring up
the sidewalks and any adjacent structures whose footings are higher than
the low point of our excavation.

Before a Structural Engineer can design the foundation system for a
building, he must know the bearing strength of the soil (or rock) of the site.
The Structural Engineer cannot wait until the excavation is completed, because the entire structure must be designed for preliminary cost evaluation
and for contractors' bidding. To discover the nature of the subsoil and the
depths where there are strata changes, the designer retains a testing laboratory to make core borings into the soil (and perhaps rock) in the area of the
intended structure. With this information the Structural Engineer will be
able to design his foundations, and the boring diagrams will give information to bidding excavation contractors so that they can calculate their pro-
jected costs. The Architect-Engineer (A/E) usually covers the owner (and
himself) with the statement that the boring charts given bidding contractors
"are for *guidance only* and that the owner and the A/E do not attest to their
accuracy." Regardless, the borings are usually a most effective guide if the
owner has allowed the A/E to order enough borings. Just as the A/E has
covered himself and the owner with a stipulating clause, an excavating
contractor who is bidding the job will usually protect his own interests by
making his total bid subject to extra reimbursement for rock that is over 1
cubic yard in size or volume. Such a clause is most necessary in glacial areas
where, even though the majority of the excavation could be indicated as
loose material, there will be a number of very large boulders on the site that
could be costly to remove.

We noted in the first part of this section that the excavator must insure
that loose material does not cave in and that adjacent structures are

shored. In a sandy condition, both requirements would necessitate extensive shoring work, but the excavation itself would be relatively easy. Conversely, an excavation into hard or semihard rock and the required blasting would be costly, but there would be little need for shoring of adjacent structures and peripheral sidewalks. All these items go into the excavation contractor's bid and into the Project Superintendent's safety thinking.

Shoring is most certainly needed to protect footings of adjacent buildings that are near to the new excavation. But what is "near"? From the study of **soil mechanics** one learns that an unshored earth bank will assume an *angle of repose*. That is, if you excavated into a bank of fine sand, you might expect the sand on the four sides of the excavation to slope to approximately 60° from vertical. However, if you excavated into solid granite, you could expect that the periphery of your excavation would remain at 0° from vertical or, in fact, the periphery of your excavation would be vertical and you might not need to shore any adjacent structures. Somewhere in between these two cases falls every excavation contractors must handle. Regardless, if shoring is necessary to protect an adjacent building or pedestrians, the Project Superintendent must know of these requirements and is responsible to his company and the community to see that the safety requirements are met by the excavation contractor or, if they are not, by his own forces.

8–2   Shoring and Underpinning

First, let us consider the shoring of the main excavation for the structure. If the building is to be far enough from sidewalks and property lines, the contractor may elect to allow enough room from the basement wall footings and then let the earth slope back gradually. However, this approach mandates that concrete trucks (for certain pours) stay farther from the building line, and that concrete for the footings and peripheral basement walls be poured from a crane bucket or from concrete buggies. Thus the General Contractor or the Construction Manager must decide how this concrete is to be poured before invitations to bid are sent to excavation contractors.

If the excavation is to be close to streets or sidewalks, or if there are reasons that the banks cannot be sloped back at the approximate angle of repose, the excavation must be shored with **sheet piling**, which consists of interlocking steel members (see Fig. 8–1) driven into the ground before the excavation starts with a pile-driving type of hammer or with **soldier beams** and **walers**. Soldier beams are steel H-beams driven into the soil every 4 to 10 ft before the excavation starts; thereafter, walers (horizontal timbers, usually 3- by 12-in. plank) are placed between the H-beams to retain the earth. They are called beams because they are *vertical* beams (i.e., cantilever beams), and they are called "soldier" beams because they stand in lines like soldiers (see Fig. 8–2).

Figure 8–1. Sheet piling. (*Courtesy of Bethlehem Steel Corp.*)

Figure 8–2. Solder beams.

The second portion of shoring to be considered is that of adjacent structures. Primarily, an excavator (or his specialty shoring contractor) must support buildings whose footings are higher than the bottom of his excavation; even if the bottom of the footings of the adjacent structure are not much higher than his bottom grade, they must be protected from "sliding" toward his new excavation. Shoring of an adjacent building and its footings is usually accomplished by digging the excavation little by little in spaced-out segments. At each segment a new footing is poured and, thereafter, a concrete wall or pier is poured to support the higher structure (or footing). This concrete wall, footing, or pier is poured within 2 in. of the structure (or footing) that is to be supported. The next day (after the first pour has taken its initial setting and shrinkage) the last 2 in. of concrete is dry packed or placed with an expansion additive so that there will be no shrinkage and there will be complete support. After these segments are strong enough to take the load, the excavations for the segments in between are made and the structure between is supported as before.

In addition to supporting the foundation or the footings of an adjacent building, the excavation contractor or his specialty shoring contractor may have to protect the masonry walls of an adjacent structure. This is done, most usually, by bolting steel channels (with through bolts) vertically up the masonry.

8–3   Cuts and Fills

Excavation for a highway involves **cuts** and **fills.** If one were grading for a highway, he would take the tops off the high spots on the route of the highway and would transport this material into the low spots on the route. Of course, the material placed in the "fill" areas would be compacted in layers (usually by a "sheepsfoot" roller) so that, when highway paving was eventually laid, there would be no settlement.

Whereas compaction of "fill" under basement slabs is acceptable (this compaction is usually accomplished with gasoline tampers or vibrating compaction equipment), it is rarely allowed under the **actual footings.**[1] If excavation has gone too low, the concrete footing must go down to virgin (undisturbed) substrata. Thus, if the elevation of the substrata was correct but the top surface became frozen the night before pouring of the footing, this frozen material would have to be removed (to virgin, unfrozen, strata) and the footing poured deeper. If this precaution were not taken, the footing would settle under load when the soil thawed. Therefore, it is important that the subgrade be protected with hay or heat until it is covered by the poured-concrete footing.

8–4   Soft Rock and Hard Rock

We noted in Section 8–1 that the excavation contract would, most probably, have a **rock clause,** which usually states that, if boulders or rock over 1 yd³ are found during the excavation, the excavation contractor will receive an "extra" of X dollars per cubic foot. One might feel that an excavation contractor should take his chances and submit an all-encompassing bid. However, if the bidding contractors took this tack, all the bids would probably be too high. The owner is better served if he accepts a lower bid that does *not* include rock (unless the borings indicated considerable rock that can be figured by the bidder), but includes an extra for large-volume rock that is not exactly indicated by the borings.

If the excavation contract contains a rock clause, the Field Engineer must measure all large boulders and/or rock strata and must keep accurate records so that the excavation contractor's requisitions may be checked. Similarly, if an entire strata in the excavation is rock, levels must be taken so that the volume of rock excavation can be calculated and entered into the records.

Although there is soft rock such as serpentine and loose shale that can be excavated with a power shovel more easily than clay (which has a great suction factor), this soft rock usually falls into the same contractual category as any rock, even the hardest. However, if the contract calls for different compensation for different rock, the Field Engineer must keep additional records.

---

[1] The author does know of a few instances where, when spread footings were calculated for 2 tons per foot (a value well-compacted sand can supply), compacted (and compaction-tested) sand fill was allowed. However, this procedure is most rare and is not generally recommended.

Throughout our text we have mentioned safety precautions that a General Contractor and his subcontractors must plan and implement. The excavation phase of a project has the same safety requirements. The periphery of the main excavation must be adequately protected from caving to protect construction personnel, peripheral construction, and the public. Excavation ditches for piping must be sheeted (when more than 4 ft deep) to protect those working in the ditch. Safety precautions on a project are "Safety Everywhere!"

Chapter 9

Foundations: Spread Footings, Caissons, and Piles

From the beginning of time, almost, many jokes have been made about building the penthouse or upper parts of the building before placing the foundations. Unfortunately, some construction companies do build certain portions of a structure out of sequence. However, anyone should know that the building starts with the foundations; in this chapter we shall discuss the different types of foundations and how supervisors should watch over their construction.

Actually, however, in the *design* of a structure, the Structural Engineer must find the load of the penthouse—snow load on the roofs—and all loadings of upper floors (including wind loads) before he can design the foundations. He must total these loadings before he can design his foundations. As mentioned in Chapter 8, the Structural Engineer must evaluate the load-bearing ability of the soil below his structure before he decides what type of foundation he must use; for this decision, he will use the results of borings or test excavations.

The foundation-supporting material under the site of a new structure may vary from solid rock, to "hardpan" (which is a combination of sand, clay, and small rocks), to what the detergent commercials term "gooey mud." If the core borings show good rock a few feet lower than the basement of the new structure, the designing Structural Engineer has few problems. He will use this rock to support the structure and will design **spread footings** to transmit the load of each column (which in turn supports a portion of the entire building's load) from the column to the rock below.

"Spread" means that the area of the footing must be expanded to give an area (of concrete footing) that will be sufficient to spread the **point load** of the column to the **per square foot loading** that the substrata will sustain. (See Fig. 9–1.)

In most cases there is a spread footing for *each* column loading, except that basement or peripheral foundation walls that support the exterior usually have one continuous spread footing. However, under some soil conditions, the design engineer elects to support all the columns (including the exterior basement or peripheral walls and their columns) on one large spread footing for the entire structure. This is called a **matte.** A matte has

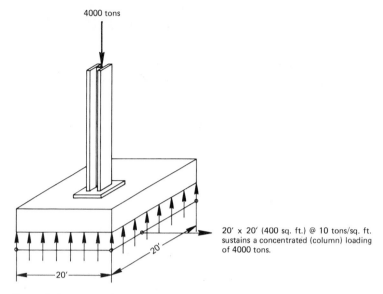

4000 tons

20′ x 20′ (400 sq. ft.) @ 10 tons/sq. ft.
sustains a concentrated (column) loading
of 4000 tons.

Figure 9–1. Typical spread footing system.

the advantage that, if there is a minor settlement condition in one area of the foundation, the matte will average out the loadings, and any settlement will tend to be uniform for the whole structure. The matte system also has the advantage that, if there is a water condition, it will be minimized in that there is one continuous footing. There are the disadvantages that excavation for foundations has to be more extensive, reinforcing steel has to be continuous, and pour joints (which we shall discuss in Chapter 17) must be carefully made unless the entire matte can be poured at once. Single spread footings (called **pile caps**) are used to consolidate the bearing value of piles, as will be covered in Section 9–6. Often one single pile cap is poured over all piles, which acts very much as does a matte spread footing that bears onto soil.

9–1 Spread Footings

Perhaps the most interesting explanation of load spreading is a discussion of the spike heels that women found fashionable some years ago. These nylon-capped heels on women's shoes were approximately ¼ in.² (0.062 in.² in area). Thus a petite 96-lb girl could exert a load of 1536 lb per in.² onto floor tiles as she walked. The result was many, many dents in the toughest of resilient floor coverings. The British government took the problem so seriously that it encouraged women working in government service to be less fashionable and wear shoes with heels of larger area. Thus, even if the worker were twice as heavy as our petite example (say 192 lb), she would exert but 160 lb per in.² if she wore a shoe with a heel with a 1.2-in.² area. The pendulum of fashion has swung and, currently, women's shoes are more sensible. Thus the ankles of ladies and resilient floor coverings are safer. This example should help us to understand the principles of building loads.

All the loadings of a building must be sustained by the footings of each and every column of the building. These column loadings will include the weight of the roof and the snow-weight load on the roof (in cold areas), and will include the weight of every floor in the building and the loads supported by each of these floors. On the structural design plans submitted to the Department of Buildings, these loads will be broken down into *dead* loads (the weight of the structure) and *live* loads (the furniture, equipment, and people that each floor must sustain). In addition to regular vertical loadings, loadings caused by wind pressures against the building must be included in the structural design.

If the bottom of a column must support a load of 2 million lb (2000 **kips** in engineering terms, or 1000 tons), and the stratum[1] that is supporting the building can support 40 tons per ft^2, then this 1000-ton column loading must be spread onto 25 ft^2 of the supporting rock strata. A 5-ft^2 concrete pad (sufficiently thick to resist cracking) with either reinforcing steel or thicker concrete (as the designer elects) would supply the necessary transfer between the column and the ground.

"Hardpan" (as previously defined), which is overlying rock, would sustain 9 tons/ft^2. Thus, if hardpan were the base of foundations, the Structural Engineer would spread the column load onto 112 ft^2 (i.e., $1000 \div 9$) and would show a footing of equal or better value (say 12 ft 6 in. by 10 ft). If the bearing value of the available strata becomes lower, the area of the spread footing must become larger.

In this regard, the Project Superintendent must remember that he should not allow the first footing to be poured until the Design Engineer (or Engineer of Record) has viewed the soil at the footing level to be sure that it is the type of material upon which his design is based. As excavation proceeds, it is the Project Superintendent's duty to notify the Design Engineer when new excavation reveals a base material that is apparently different than the supporting stratum that was previously approved. If the newly discovered soil has less supporting value than the expected soil, the Design Engineer might decide to require a footing with a larger area or to excavate deeper to better strata.

There are times when the supporting strata are considerably lower than the elevation of the cellar or basement of the structure. In such cases the Design Engineer may elect to carry the load to this lower supporting strata by means of "Caissons" or "Piles."

9–2   Caissons

Originally, **caissons** were large, boxlike, structures that were built to retain (or keep out) mud or water and were set into river banks in order to excavate for bridge foundations. As the excavation proceeded inside these

---

[1] Stratum is the natural layer of supporting earth or rock.

caissons, they sank or were built down, and, when the required bearing strata were reached, a concrete footing was poured and a masonry bridge pier was built. When the foundation had been completed, the caisson forms were removed, except when they became an integral part of the bridge pier.

However, for the caissons in more conventional building construction, a square or circular barrier is placed as the excavation proceeds. If the depth of the hole is not too great, the entire (shored) excavation is filled with concrete (and steel reinforcing when required) and the shoring is either pulled or abandoned. A caisson of this type is shown in Figure 9-2(a). The area of the bottom of the caisson is the area that will support the loading of the column (and the weight of the caisson's concrete).

In many cases the required bearing strata are so far below the building elevation and the required bearing area is so large that filling the entire full-sized caisson shaft with concrete would be economically prohibitive. In such cases a circular excavation is made (or drilled), and a steel shell is lowered (when required to withstand unstable soils) as the drilling or excavation proceeds. When the desired (design) strata are reached, the bottom of the excavation is **belled out** to provide an area that will support the load. Originally, these excavations were made by men who were lowered into the hole (or casing) and sent the material to the surface in buckets. Now, however, the usual procedure is a machine process that brings the excavated material to the surface and, when the bottom soil has been approved by the engineer's representative, bells out the bottom. After completion of the bell and the final inspection, required reinforcing is placed and the circular excavation is filled with concrete. Whenever possible any steel casing is withdrawn as the concrete is poured. Thus we end up with a "spread footing" at a lower depth and a circular concrete "column" that supports one of the building's columns. A belled caisson is also shown in Figure 9-2(b).

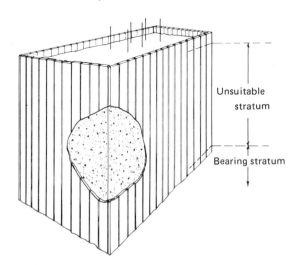

Unsuitable
stratum

Bearing stratum

Figure 9–2.   (a) Regular caisson (plank-sheeted interior bracing omitted).

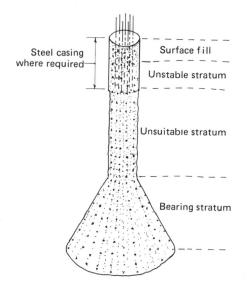

Steel casing
where required

Surface fill

Unstable stratum

Unsuitable stratum

Bearing stratum

Figure 9–2. *(Continued.)* (b) Belled caisson.

9–3   Drilled-In-Caissons

When solid rock is available at a considerably lower depth, the Design Engineer may elect to go for this more substantial support because higher strata do not give what is required to keep the size of the footing within reasonable bounds. In such cases a casing of heavier gauge than used for the belled caisson is required. This casing will have to be thicker because it must withstand the (horizontal) pressures of deeper mud and water, and because the thickness of this casing or shell can be used for support (i.e., the area of steel shell, plus the area of concrete, plus the area of steel reinforcing inserted are all combined into the final load capacity of the drilled-in caisson). This excavation will be made very much as a water well is excavated; when rock is reached, drill bits will be used to make a hole (or socket) into the final rock strata. (Note that if minor rock is encountered at higher elevation a rock bit must also be used here.) After the desired rock strata are encountered, this heavier casing will be filled with concrete. However, because the bearing value of the rock strata is great, this shaft can support great loadings if the concrete shaft is reinforced with a large amount of steel. Thus a heavyweight "H column" is usually dropped into the casing (each section welded together as was the original casing), centered, and then concrete is poured. The combined strength of the heavy steel casing, the heavy steel H column core, and the concrete filling provides a very strong support and is often used when high, concentrated loads require support and excellent bearing value is available at lower elevations.

This system, termed the "Drilled-in-Caisson," was originated by H. Thornley of Western Foundation Corporation. It is used extensively when high column loadings are necessary and when high-bearing-value strata are available at lower depths. This system has the advantage that, like piles (discussed next), the casings can be **battered** (i.e., driven at an angle) to resist horizontal loads in addition to the usual vertical loads. A typical illustration of the drilled-in-caisson system is shown in Figure 9–3.

ROCK-SOCKETED   DRILLED-IN-CAISSON

INSTALLATION PROCEDURE

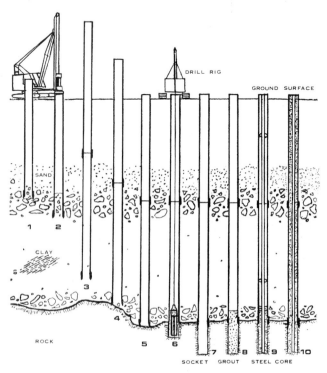

Figure 9–3. *(Courtesy of Western Foundation Co.)*

9–4   Piles

Often soil conditions (as revealed by test borings) do not afford the bearing value at higher elevations for spread footings, and at lower elevations do not afford the conditions required for caissons. When confronted by such a problem, the Structural Engineer may decide on pile foundations.

Piles have been used for foundations for centuries. Originally, they were used for bridges and waterfront structures such as wharves. The piles were wood shafts (straight sections of trees). In later years wooden piles were treated with creosote or other chemicals to prevent rotting and to protect them from boring worms.

The bearing capacity of a pile is dependent on two things: (1) the strength of the pile itself, and (2) the supporting value of the soil into which the pile is driven. Thus, if a concrete or steel shaft of considerable strength were driven through mud, silt, or whatever, until the tip of the pile reached hard rock or strata that were stronger than the pile, which would not allow further penetration (called "refusal"), the support value of the pile would depend only on the strength of the pile shaft itself. In such cases the Design Engineer could choose a very strong H column or use a pile with (say) 5000-pound per in.^{2} (psi) concrete. However, if the pile depends on less than refusal conditions, it does not make much sense to spend money to have a shaft greatly stronger than the bearing value that the earth or substrata can provide.

Pile support is achieved by (1) tip support, which is usually "refusal" conditions, (2) "skin friction" or the support provided by the cling of the surrounding earth, or (3) by a combination of both. Therefore, depending on the support available, the Design Engineer must choose a type of pile and strata for support.

WOOD PILES

Wood piles, the original piles, are limited by shape. It is fairly difficult to find a straight tree longer than 60 ft. In addition, the tip size of such a tree is rarely more than 6 in. Thus, whereas loading on wood piles is allowable to 30 tons, the shape of the piles available usually limits the loading value to 20 tons each. Originally, wood piles were used as skin-friction piles for loadings of 15 tons. However, now that greater design loadings are allowed by codes, they are used for 20 tons and, when shape allows, to 30 tons.

9–5 Bearing-Value Calculation and Pile Tests

H piles, precast concrete piles, and piles composed of sheet metal or pile casings can be manufactured in any size or strength and can, therefore, achieve bearing value equal to the strata available. The engineer and the pile-driving team can *calculate* the bearing value of a pile by one of several formulas. Regardless, after one or more piles have been driven, test loads must be placed upon one or more of the piles that may be considered typical for the project to *prove* the bearing value of the piles driven through and to similar strata. One of the older of the formulas, and one cited in the codes of many cities, is the **Engineering News formula:**

$$R = \frac{2\,WH}{s + 0.1}$$

In this particular formula (for single-acting hammers),

R = allowable pile loading, in pounds

W = weight of the hammer (i.e., the moving part), in pounds

H = effective fall distance of the hammer, in feet

s = net penetration in inches per blow for the last five blows after the pile has been driven to a depth where successive blows produce equal net penetration

Thus capacity is a function of *inches of pile penetration per blow* of the pile-driving hammer.

This formula is what we call an "empirical" formula; that is, it is a formula arrived upon *after* the problem was solved. The formula was then written to conform with the experience of many tests that proved the system and conformed with the *results,* rather than calculated first to figure what results would be achieved.

PILE TESTS

As previously noted, after a pile (or piles) is driven in accordance with a formula, the load-capacity must be **proved.** The proof is really what the designer and a municipality's Department of Buildings desire. In most localities where the Engineering News formula is a part of the building code, a pile is tested by placing *twice* the design load onto the pile. After a pile has been driven in accordance with a formula and has achieved the rating, required loads are placed onto it in increments. If the pile is a concrete-filled casing pile, it must be allowed to set (unloaded) until the concrete has achieved sufficient strength. Thereafter, a crib is set onto the top of the pile and a load is added (usually in seven increments) until the load is twice the design capacity of the pile. Grade readings are taken after each load increment to an accuracy of $\frac{1}{1000}$ ft. After this test (twice the design loading) is fully applied and there is no settlement within a 2-hour period, the load is kept on the crib until any (further) settlement does not exceed $\frac{1}{1000}$ ft in 48 hours. When this condition is achieved, the test load is removed in not less than four increments, and, at each reduction of load, the elevation of the pile is read and recorded. Under ideal conditions the pile should elongate (or rebound), and the final (top) elevation should equal the original (unloaded) elevation. This rarely happens, but, if reasonably close, the pile and the system are acceptable. Under conditions set by many codes, the maximum pile load shall be one-half that which causes a net settlement of not more than $\frac{1}{100}$ in. per ton of the total test load or shall be one-half of that load which causes a gross settlement of 1 in., whichever is less.

The previously described pile-resistance formula and pile test are only examples. Originally, these were the only acceptable systems. However, as the pile-driving system has been improved with new and improved driving hammers and as new pile casings have been developed, greater loadings have been made possible. Thus the reader will understand that new formulas have been developed. For example, one knowledgeable group of State engineers has added a calculation on the original Engineering News formula to take account of the impact-loss factor of driving systems. A noted foundation engineer has evolved a formula based upon wave equation which includes the values of the cross-sectional area of the pile, the modulus of elasticity of pile material, and the length of the pile segment. In addition, there are formulas for piles driven by sonic methods. All these formulas have their place in the particular pile-driving system. But, basically, it is the test that is important.

Any pile-driving specification written by a design engineer for a particular foundation system is a performance specification. Whereas this engineer may specify the type of pile he feels is appropriate for the soils indicated by borings, the final result is his requirement for a certain number of tons for each pile driven. This capacity must be proved by the testing of one or more typical piles. Here too there have been new innovations. We have described the original and basic method of pile test because it is simple and easily understood, but there are a number of different test methods. One of the most prevalent tests uses a load applied with hydraulic jacks acting against a reaction frame that is held down by anchor piles. Regardless, the formula and the method of testing may change, but the final result must be a pile that tests up to the values required by the Design Engineer and the requirements of the local Department of Buildings.

Figure 9–4 shows a pile driver, a typical piece of equipment.

Figure 9–4. Pile-driving rig. (Courtesy of Raymond International, Inc.)

9–6   Types of Piles

Piles can be wood, steel H sections, precast concrete shafts, sheet-metal casings filled with concrete after driving, or pipe-section piles which are filled with concrete after driving. Under most usual conditions a number of piles (each of the desired capacity) are driven in groups (or **clusters**) for each column. For example, if a column in a structure required piles with a combined capacity of 950,000 to 960,000 lb (475 to 480 tons), the designer might provide six 80-ton piles in a cluster. There are several ways in which he might arrange these piles. Two possible arrangements are shown in Figure 9–5(a) and (b) along with vertical sections showing how these piles would extend (usually 4 in.) up into the **pile cap.** The reader will understand that, when groups of piles are used, a concrete cap is necessary to consolidate the bearing capacities of all piles in each cluster or, if you will, to distribute the load of a column onto all the piles in the cluster.

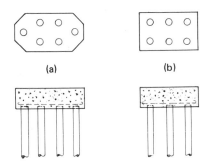

(a)   (b)

Figure 9–5.   Two possible arrangements of piles are shown in (a) and (b).

No one type of pile **is best** for all projects. Every project has special needs. Subsoil conditions vary. The Design Engineer must choose the pile he feels will serve the project best. This text will describe the potentials of the different types of piles. However, when bearing values are mentioned, the values are only *relative.* As construction costs rise and construction procedures are perfected so that better and more uniform results may be achieved, safety factors may be reduced and engineers may use a greater portion of a pile's potential. *Each month* there are developments in pile-driving machinery and technique that are increasing the bearing capacities of piles. For example, concrete-filled sheet-metal casing piles that were usually driven to 50-ton capacities in the 1950s were being driven to 80-ton capacities in the 1960s; currently, there are piles of this type (with heavier steel tip sections) that have been driven to capacities in excess of 200 and 300 tons.

Figure 9–6(a) through (d) shows cross-sectional diagrams of four sets of pile system. These diagrams are drawn to show shapes and are, thus, exaggerated; they are not to scale. Figure 9–6(a) shows a wood-pile cluster. Because of the basic shape and tapering of a wood pile, it is usually used as a "skin-friction" pile and is usually used in systems that requre 10- to 20-ton piles.

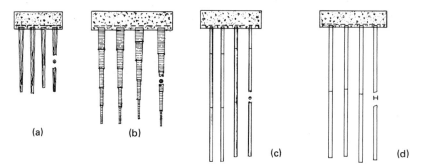

(a)   (b)   (c)   (d)

Types of
Piles

Figure 9–6. (a) A wood-pile cluster. (b) A Step-Taper pile cluster. (c) A pipe-casing pile cluster. (d) A cluster of H-beam sections.

Figure 9–6(b) shows a Step-Taper pile cluster, which was originally invented by engineers of the Raymond Concrete Pile Company (now Raymond International, Inc.).[2] This pile casing is made up of corrugated sections that screw together to make up the required pile length. The corrugation allows a lighter-gauge metal to be used, and the corrugations (combined with the taper of the pile) give considerable skin friction. Because of the relative thinness of the metal casing, this type of pile casing is driven by inserting a solid-steel driving mandrel that exactly fits the tapering contours of the casing, including shoulders on the mandrel that match the drive rings on the casings (see Fig. 9–7). This mandrel accepts the thrusts of the pile-driving hammer and pushes the pile into the ground. After the casing is driven to design resistance, the driving mandrel is removed and the casing (along with other casings driven in the same cluster) is inspected to be sure it is clear of water or mud (which could be admitted by a tear made by a subterranean obstruction); the casing is filled with concrete whose design

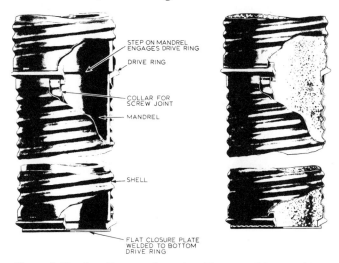

STEP ON MANDREL
ENGAGES DRIVE RING

DRIVE RING

COLLAR FOR
SCREW JOINT

MANDREL

SHELL

FLAT CLOSURE PLATE
WELDED TO BOTTOM
DRIVE RING

Figure 9–7. Step-Taper configuration. (*Courtesy of Raymond International, Inc.*)

[2] Step-Taper is a registered trade name of Raymond International, Inc.

strength is formulated to work with the required loading of the pile. In addition, the Step-Taper casing may be combined with pipe tips to give further drive length. One should note here that there are other sheet-metal casing piles available from different manufacturers.

Figure 9–6(c) shows a pipe-casing, concrete-filled pile cluster. Pipe casings have much thicker walls than sheet-metal casings and, therefore, no driving mandrel is required. After a steel cap end is welded onto the bottom end of the first section, the pipe casing is driven into the ground. Other pipe casings are welded to it until it reaches the strata and resistance required by the design. Then, after inspection for possible damage shows that the interior is clear, the casing is filled with concrete. Although this pipe-casing pile does not have as much skin friction as a tapered section would have in similar strata, it does have considerable skin friction. However, because of its casing strength, it is often chosen for tip loading. Depending on strata and use, this type of pile is used for loadings of 50 tons and upward.

Figure 9–6(d) shows a cluster made up of H-beam sections. H beams can be chosen for many strengths. When one figures the length of the total periphery of an H section as opposed to the periphery of a circular pile, one realizes that this section has considerable area for skin friction under certain subsoil conditions. However, because of the strength of this shape, the H pile is usually used for tip loading.

Not shown in the illustration are prestressed, precast concrete piles. These are precast to lengths required by the particular foundation design and test, and are driven by the conventional methods previously described.

9–7 Pressure-Injected Footings

Pressure-injected footings are often termed "piles" because the supporting concrete is driven into the ground. However, this is a misnomer. Actually, this type of footing is achieved by "forging" a rather large ball of concrete within granular strata, which has a column of concrete extending up from the ball footing to the surface. This type of footing requires a soil that has granular characteristics. A soil with considerable clay would not be suitable. The originator of this system was Edgar Frankignoul, who established a company called Franki. His invention was called the Franki Systems, but many engineers called it a Franki pile. Later, other engineers called it a **bulb pile.** These days, several foundation companies that install this type of footing have their own names for it. But it is not a pile. Engineering specifiers term it a pressure-injected footing.

A pressure-injected footing is illustrated in Figure 9–8. A true pressure-injected footing starts when a steel casing (say 20 in. in diameter) is placed in position. A plug of gravel (approximately 5 ft³) or zero-slump concrete is placed inside the casing. Then a very heavy drop hammer is lowered onto this plug. We should note that the diameter of the drop hammer is smaller than the casing and that it is dropped from 10 to 20 ft for each blow. As the gravel or concrete is driven into the soil, it arches and drags the casing with

INSTALLATION PROCEDURE

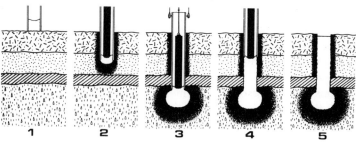

Shafts are cased where intermediate strata requires

Figure 9–8. A pressure-injected footing. (*Courtesy of Franki Foundation Co.*)

it. At a predetermined design depth (here, again, test borings are utilized), the casing is supported by the "leads" of the driving rig so that it will not descend further. Then small quantities of zero-slump concrete are dropped past the drop hammer and the driving continues. Little by little more zero-slump concrete is driven into the strata by the hammer. Because there is now no inhibiting casing, this concrete expands laterally into a "ball" the size of which increases as the amount of concrete is increased. When the ball footing is the correct size, more concrete is added to the casing, and the casing is raised, foot by foot, as the concrete is added. At the top, reinforcing cages and anchor bolts are inserted as required by the design of the connecting structure. This system, as herein described, can afford support of 80 to 200 tons. Uncased shafts are not recommended through soft clays, silts, peat, or other organic soils.

The limiting, practical depth of the pressure-injected footing is about 80 ft. The problem with pressure-injected systems is that the soil (at the bulb level) must be granular or naturally resistant, and the concrete must be of very low slump. If the substrata (where the "ball footing" is desired) are not granular or highly competent, or if the concrete is not low slump, the ball may not be properly formed (i.e., it might be formed to one side or eccentric to the main shaft), and the off-center footing might exert bending forces on the shaft of the footing. However, in this regard, a reputable foundation contractor who specializes in this type of footing would be the first to advise if soil conditions would not permit his work to be acceptable to the design.

9–8   Responsibility for Bearing Value

The Project Manager and the Project Superintendent must be aware of the responsibility involved in the foundation work covered in this chapter. As previously noted in the discussion of spread footings, which are designed by the Engineer of Record to set onto soil strata or rock, the Project Superintendent (or his company) is responsible to advise the Engineer of Record when the bearing strata are reached and are ready for inspection; the Engineer of Record will then decide if the strata are as good as he anticipated. If the strata (which he previously approved) changes as the excavation proceeds, he should be requested to revisit the project and recheck the new areas. However, it is his responsibility to check the soil, inasmuch as he designed the structure and placed his seal and signature on the design drawings.

When pile foundations (and caissons) are involved, the Engineer of Record is responsible for the design, and the pile (or caisson) contractor is responsible for inserting the piles (or caissons) with the correct resistance; but it is the *owner's* Professional Engineer's responsibility to supervise and attest that correct results have been reached for each pile or caisson. One might feel that the Designing Professional Engineer should watch the driving of piles or the placing of caissons and attest to the results. And, for his own satisfaction, he does supervise a portion of this work. However, under normal conditions, the specifications will state that the owner shall retain a Professional Engineer who specializes in this type of foundation work to observe the work and attest that each pile (or caisson) is correctly placed. He will also sign Department of Buildings documents that design conditions have been reached for each pile or caisson.

We conjecture that this usual requirement is intended to remove all possible conflict of interest. A shoddy contractor or pile-driving subcontractor might wish to accept "almost" correct results to speed the project. The Design Engineer might wish to be absolutely sure and go for a few more "blow counts" on the piles (this might be costly). A Professional Engineer (and a specialist in pile driving and caissons) is there to assure without a doubt that the interests of the owner and the Department of Buildings are served. This endeavor is usually shared by all concerned. However, the responsibility is usually placed upon a third party.

Chapter 10

Peripheral (Cellar) Foundation Walls and Their Protection from Water

The cellar or basement of a building is a most important space. It houses electrical switchgear, it may house air-conditioning machinery, and it may even house office space. None of these functions operate efficiently or, in the case of personnel offices, happily, if there is leakage into the sub-street-level spaces. Therefore, in addition to discussing the method of pouring the peripheral walls, we shall discuss methods of keeping water out of the basement or cellar.

10–1   The Peripheral Footing

If the footing system of a building is a spread-footing system (see Chapter 9), there will be a continuous spread footing under the exterior foundation wall or the exterior (peripheral) foundation walls of the building. If the foundation of the building is a pile system, there will be strings of piles around the periphery of the building, which will have a continuous pile cap. Regardless of whether the system is a pile cap or a spread footing, it will look like the footing shown in Figure 10–1. The exterior walls of the basement or cellar will sit on this footing and will support the exterior columns of the building.

If no water problems are indicated by the records of the area or by the test borings that were taken for foundational purposes, the designer will make no provisions for water exclusion, but will, regardless, set a "key" which is formed by setting a tapered two by four into the wet footing concrete so that the bottom of the foundation wall will have a resistance to the inward push (or **shear**) from the exterior earth. In addition, there will be reinforcing rods coming out of the footing for the wall. These will add shear resistance. If there is a water problem, the designer will call for polyvinyl chloride (PVC) waterstops between the peripheral footing and the bottom of the peripheral wall and between the side of the peripheral foundation wall and the cellar floor. In addition, he will call for waterstops between wall pours and between floor pours. Such waterstops are illustrated in Figure 10–2.

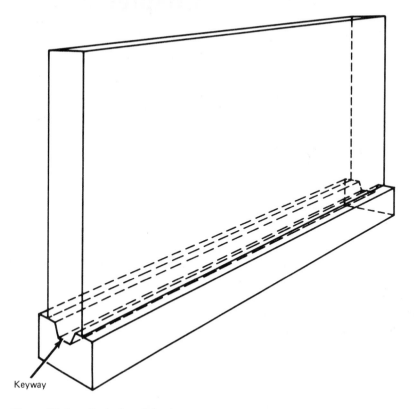

Keyway

Figure 10–1. Typical wall-footing system.

Before we discuss the forming and pouring of these peripheral walls and the setting of PVC waterstops, we shall list the ways by which water could enter the basement:

1. Groundwater may enter into the basement of a building through the construction joints (and key) between the footing pour and the pour of peripheral wall.

2. Groundwater may enter into the basement through the construction joints (or pour stops) between sections of the peripheral wall.

3. Groundwater may enter through construction joints between the peripheral foundation wall and the cellar slab pour.

4. Groundwater may enter through pour joints (or pour stops) in the basement floor pours.

5. If the building (above the basement level) steps back (i.e., has a setback from the building line) from the line of the peripheral basement walls, a leak through the sidewalk slab that covers a basement below could pass through this slab and fall onto equipment or desks below.

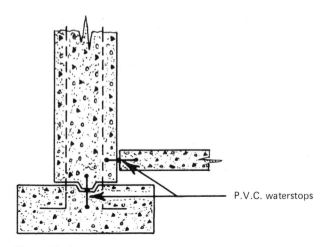

Figure 10–2.

6. Groundwater may pass directly through basement walls where concrete was not sufficiently dense.

7. Groundwater may pass directly up through basement floors if there is sufficient head or if the subsoil below the concrete slab is not porous.

Figure 10–3 shows a portion of a basement wall sitting on a keyed continuous spread footing. The diagram shows a PVC waterstop between the footing and the concrete foundation wall, and a PVC waterstop between the concrete foundation wall and the concrete floor slab. Note also that the soil under the floor slab is covered by at least 6 in. of granular fill, which has a 6-mil vapor barrier (i.e., a 6-mil polyethylene sheet that should have taped connection joints) between the granular fill and the bottom of the concrete. The granular fill is placed there to drain away any moisture that might accumulate, and the vapor barrier is placed between the granular fill and

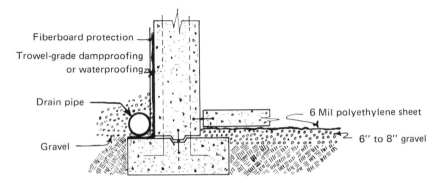

Figure 10–3.

the bottom of the basement slab to prevent any moisture from making the basement floor damp. At this point we should explain that, if the granular fill is drained to a lower point, there would be no water pressure pushing water up through the slab. However, the slab itself might absorb water, which would tend to create dampness on the top surface of the basement floor. Also, if the designer feels that there is a water-pressure problem, he will place terra cotta, concrete, or fiber drain piping around the outside of the basement walls (as also shown in Fig. 10–3), which will be pitched to a draining area. Thus, even if there is water next to the substructure, it should drain away from the basement walls and basement floors to a lower point of the site.

Finally, we advise that a **trowel-grade** dampproofing compound, such as an asphalt cut-back or an asphalt emulsion, be applied to the exterior of any basement wall and onto the joint between the footing and the basement wall. If there is a more prominent water condition, the designers will call for greater dampproofing procedures. However, our point here is that, while the exterior of a basement wall is uncovered, it is folly not to put some kind of dampproofing on it. To protect this trowel-grade dampproofing, sheets of asphalt-impregnated fiberboard should be placed against the foundation lest rocks or other hard objects in the backfill make penetrations in the mastic coating.

10–2   Condensation

A moisture condition that should be considered before basement walls are backfilled is caused by moisture *inside* the building and weather conditions *outside* the building. If the basement or cellar activities are such that there is warm air in the basement in winter months where the outside of the basement wall will contact very cold earth, the basement wall will be cold and will tend to condense moisture in the interior, air-conditioned space. If the interior of the basement is to have conditioned air, the designer will wish to insulate the exterior basement wall. He may do this by placing a **closed-cell** insulation (i.e., insulation that will not absorb moisture) on the outside of the wall before it is backfilled, or he may place sheets of insulation on the inside of the wall and then cover the insulation with gypsum board. If there is to be insulation on the exterior of the basement, this must be placed and protected before backfilling.

10–3   Moisture Through Street-Level Slab

A number of our newer office buildings are taking a large setback at the street level for artistic or zoning reasons. This leaves a sidewalk-level promenade slab covering a considerable portion of the basement. In such cases the basement must be covered and protected just like *any* space under a roof.

There is a slight difference here, however. Any elastomeric or built-up membrane water protection (see Chapter 20) must have even more protection from traffic before it is covered with the concrete or terrazzo walking surface. Because this waterproofing is near to the path of construction men walking to and fro on their constructional duties or to their lunch, interim protection must be placed over the waterproofing until the final wearing surface is placed. Because of many damaged areas to such waterproofing, many construction superintendents feel that workers would rather walk on new waterproofing than on cured sidewalks and that some of these workers must be wearing golf shoes! Actually, it's no joke! Street-level waterproofing is indeed most difficult to protect, but it *must* be protected. Once the wearing deck is placed, it is most difficult to find and patch a leak. Originally, 1 in. of a sand–cement mixture was used to protect waterproofing until the final topping surface was poured. Quite often, these days, 4-ft² by ⅛-in.-thick bituminous protection sheets are applied onto which topping is poured later.

10–4 Metallic Waterproofing

There are many, many systems to keep moisture from penetrating through elevator pits, sump pits, basement floors, and basement walls. The system most relied upon when severe water conditions are anticipated is the metallic waterproofing system.

Consider an iron bolt that has been exposed to many cycles of wetting and drying, such as a bolt exposed in a wharf's piling. The diameter of such a corroded bolt is much larger than the original bolt. This expansion of corroded metals is the underlying principle of the metallic waterproofing system. If one can fill all the tiny air holes or voids in a concrete wall or floor with an uncorroded metal, the voids will be packed solid as soon as the metal corrodes.

In actual practice, the interior of a concrete wall is **bush hammered** or acid etched to remove the film of pure portland cement skin that collects on a carefully vibrated concrete wall. The bush hammering or etching (the author prefers bush hammering) uncovers all the little holes and voids. Then a mixture of finely ground gray cast iron and a catalyst to hasten the rusting process (often an ammonium compound) is diluted with water and brushed onto the wall's surface. The next day a second coating of this thin liquid is applied. As these tiny particles of cast iron rust they expand to many times their original diameter and fill the voids in the concrete wall's surface. After two coats of this mixture have dried and expanded, a mixture of fine sand, metallic compound (with catalyst), and portland cement is applied to the wall in a thin "scratch coat." Later a ¾- to 1-in. cement coating is steel troweled onto the surface to protect the waterproofing layers.

The floor of a basement is protected in a similar manner. However, in this case the concrete floor is poured low to allow for 1½-in. concrete fill.

This subfloor is raked before it hardens to give the surface an adhering quality. Before the wearing surface is applied, this roughed-up surface is bush hammered or acid etched as were wall or column surfaces. The metallic coatings are applied as described previously and then the wearing surface concrete is applied over the waterproofing layers.

10–5 Pressure Slabs

It would not do much good to waterproof a basement floor slab only to have the entire slab rise when underground water pressure lifted the slab. Under most basement or cellar conditions the floor slab is designed to withstand the traffic on *top* of the floor slab. However, if the designer wishes the floor slab waterproofed metallically, there *may* be a water-pressure problem involved. If there is, he will design the concrete floor slab for uplift loadings as well as downward (traffic) loadings. In such cases he will call for hooked reinforcing rods in all footings to catch into the floor slab and hold it down. Such a slab is called a **pressure slab.** Of course, it is the Project Superintendent's duty to see that a building is constructed in accordance with the designer's drawings. He is not required to question these design drawings (unless he feels there is a radical error), because the design is the responsibility of the A/E. Nevertheless, if he sees a considerable water condition as excavation progresses, or if he notices that the design calls for metallic waterproofing of the slab and there are no **tie-downs** (see Fig. 10–4) in the footings, he might do well to discuss the situation with the designer; he may also wish to document his concern to protect his employer in case of later problems.

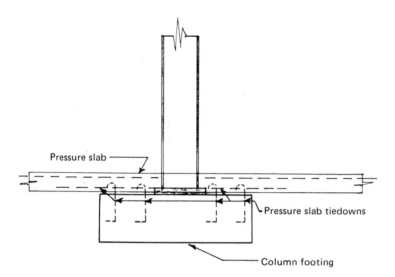

Figure 10–4. Pressure-slab tie-down system.

In the previous chapter we did not adequately discuss matte foundations. There were a number of other foundations which are derived from the systems discussed in Chapter 9 that would make for too much reading at that time. Some of these derived foundations will be discussed in this text as an accompanying subject demands. The discussion of the waterproofing of pressure slabs now requires a discussion of matte foundations.

Figure 10–5 shows several spread footings alongside each other, which is a more general system. To the right of the spread-footing system is an illustration of a matte system that might be used where certain foundation conditions require it.

A matte foundation is, in essence, one large spread footing taking the place of a number of smaller spread footings. Quite often, when the subgrade strata on which foundations must bear is of a consistency with less resistance than hardpan, or where there is a water condition that will present constant problems to the foundation contractor, the foundation slab of the building is designed as one large "spread footing" or matte as opposed to a number of smaller (spread) footings. This matte footing may have to be divided into several pours if it is too large for one, continuous pour. However, in such cases the reinforcing bars will be continuous through the pour stops (the pour stops will be set in accordance with a general system discussed in Chapter 17). This one-matte system may have an engineering advantage in that any slight settlement caused by column loads will usually be constant for the whole matte and, therefore, there will be no cracking problems in superstructure.

Because this footing is for all columns and peripheral walls, the matte will be thick and heavy. If there are pour stops in a foundation system where there is a water problem, PVC waterstops will be set between matte pours and into the edges of the matte to keep water from passing between the matte and the exterior basement wall. Of course, there will be "keys" formed in the periphery of the matte to give lateral support to the basement walls, just as one would find in a peripheral footing (as shown in Fig. 10–3).

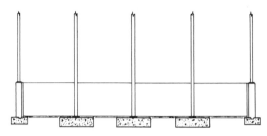

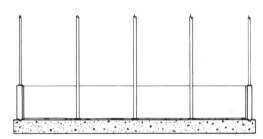

Spread Footing Foundations for a Steel Building
Steel framing omitted for clarity

Matte Foundation for a Similar Steel Building

Figure 10–5.

Here then is the reason we held back on the discussion of matte footings until we came to metallic waterproofing and pressure slabs. If there is underground water pressure, it will not lift a heavy, continuous matte. However, if the designer feels that some of the water (under pressure) could force its way through the matte eventually, he will allow for a thick fill-and-finish layer; the top of the matte is treated with metallic waterproofing in critical areas just as one would treat a thinner subslab (see Fig. 10–5). However, because of the weight of the matte, there is no worry about uplift or a need for "tie-downs."

10–7 Construction and Bracing of Wall and Column Forms

We have discussed the methods of construction that will keep moisture from passing through basement wall and floor systems. Now let's discuss better methods of forming and pouring these walls and columns.

Primarily, forms for walls and columns must be constructed to withstand the pressure of liquid concrete. Concrete weighs approximately 150 lb per ft^3. Thus, when a 3-ft deep layer of concrete is poured into a form, the pressure on the bottom of the form is 450 lb per ft^2 of formwork at the lower portion of the pour. As the concrete hardens, it sustains its own weight, and then the horizontal pressure ceases. Thus, for many years, concrete superintendents have endeavored to pour at a *maximum* rate of, approximately, 3 vertical ft per hour. In this way the lower concrete becomes semihard (thus self-sustaining), and as the fourth foot of concrete is poured, there is no lateral pressure from the first foot of concrete poured. This system worked fine until design engineers found it necessary to use higher-strength concrete mixes. Because they contained more portland cement per cubic yard, the concrete **set up** (or hardened) faster. Therefore, when the length of foundation walls was short and did not allow a wall pour to be stretched out during the pouring, a concrete superintendent had to pour faster than 3 ft per hour or accept **cold joints** (i.e., undesirable joints in concrete where one layer of concrete sets up before the next pouring). Thus, if he wishes to pour more depth of wet concrete per hour, he must design his wall and column forms to sustain more lateral pressure.

An illustration of wall forms is shown in Figure 10–6. This indicates plywood or steel forms and the vertical strengthening (which would be two by fours vertically on timber forms and angles or channels on steel forms), which is held together by horizontal walers and ties that hold the forms from spreading. If the forms are filled with concrete at a faster rate than 3 ft per hour, the form builder must install more vertical braces per lineal foot of forms, more walers per vertical foot of forms, and must increase the number of bolts or "snap-ties" that hold the forms from spreading under pressure. The walers for *column* forms are, usually, steel **column clamps.**

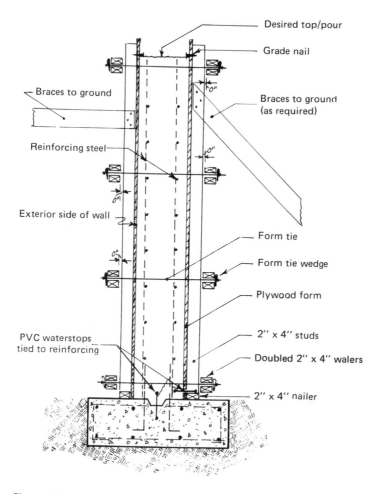

Figure 10–6. Form construction details.

Of course, there is no use pouring a wall or column unless it is braced to keep the wall plumb and linearly aligned and the column plumb in both directions. Insofar as the basement wall is concerned, these walls are kept true by braces from the ground or slab to the top of the wall form. The column forms are kept plumb by bracing them from the top of the column form to the slab below if they are to be poured before the next slab (i.e., the slab at their top) is formed and braced. If the columns are to be poured *after* the deck above is formed and braced (which is a more usual practice), the lateral bracing to the exterior (or peripheral) columns is sufficient.

If the concrete of columns is merely sheathing of structural steel columns for fireproofing purposes (see Chapter 19), the steel column will support the minor bracing (wedges) of the concrete forms.

It would not make much sense to install an intricate electronic burglar alarm in a place of business and then not install the connecting circuits. Neither does it make much sense to install PVC waterstops in a concrete pour and not protect them from bending out of the pour division.

The stiffness of most PVC waterstop materials leads the installer astray. Most of us, feeling the inherent strength of most PVC products when they are at *room temperature,* would be led to believe that the material was strong enough to support itself in a concrete pour. This is not true. When subjected to the heat caused by the chemical action in liquid concrete, PVC waterstops may soften to putty consistency. If this is not considered and tie wires are not placed on the waterstops (usually at 2-ft intervals), the waterstop coming out of the footing will lie down on the top of the footing, and the waterstop inserted into the side of a peripheral wall will lie down against the side of the wall form. In either case the waterstops will be useless, and because the condition will not be discovered until the concrete is hard (probably never on the footing waterstop), the condition cannot be corrected by other than metallic waterproofing. However, it is quite simple to prevent such problems by supporting the PVC material every 2 ft with No. 14 wire ties to the reinforcing steel imbedded in the previous pour or set for the next pour.

Finally, one must realize that PVC waterstop material comes in rolls that are not endless. Therefore, one must splice or weld one section of PVC to the adjacent section. This may be accomplished by using a hollow PVC section and cementing a core piece to join adjacent sections (see Fig. 10–8) or by placing the accurately cut (i.e., 90°) ends together and heat fusing them together, as shown in Figure 10–7. Either way is acceptable in the construction industry. However, unless we make the waterstops continuous, it does not make much sense to take extreme care in other aspects of waterstop installation.

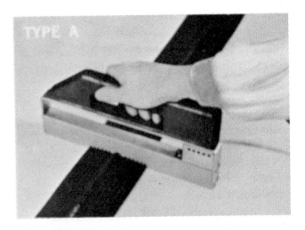

Figure 10–7.   Joining of PVC waterstop material. (*Courtesy W. R. Grace and Co.*)

Figure 10–8.
(Courtesy of W. R. Grace and Co.)

Remember always, water penetration can make a building very uncomfortable. A lot of unhappiness for the tenant and the builder can be avoided by taking great care with PVC waterstop and other waterproofing installations. Again, always "think dry." It's not all that hard to take precautions that will insure a dry basement. If there is a leak after the basement is completed, it is a major problem. Avoid this problem. Do the work right the first time!

Chapter 11

Types of Superstructure Construction

In Chapters 11–14 we shall discuss the four basic types of construction that a fairly large construction company will normally handle, beginning in this chapter with wall-bearing masonry. This text is written to help in the development of those who are following commercial construction, which embraces skills used from the earliest days of house construction. We cannot cover all these skills. However, as the construction man moves up through the building business from assistant to executive he must pick up the old-fashioned skills that have become a part of modern construction. We shall not discuss the construction of all-timber structures, "balloon framing," and the many specialties of all-timber construction. These systems are still used in building houses, but this text addresses commercial work. However, we shall discuss systems that have grown out of these original systems, as follows:

1. Types of superstructure construction: exterior wall-bearing masonry: Chapter 11.
2. Basic structural steel framing with concrete and masonry: Chapter 12.
3. Reinforced concrete framing, poured in place: Chapter 13.
4. Completely precast buildings: Chapter 14.

11–1 Background

Originally, all *large* construction projects (including castles in the old world) consisted of masonry walls that enclosed the building and supported the floors, interior framing, and roofs of the building. In the early days of construction this interior framing was wood. Soon, in an effort to achieve larger spans between supporting masonry-bearing walls or columns, some of the wooden girders were supplanted by steel **I beams;** soon after, wooden columns and stone piers were replaced with steel columns. Still, this was basically the original "wall-bearing" type of construction.

As the steel industry continued to perfect steel beams and steel columns, it was able to supply steel columns, girders, and beams, along with connection systems that could be used to provide the entire framing of a building above the foundations. In a few cases the floors of these buildings were wood. Usually, however, the floor systems were reinforced concrete, as was the roof deck. In the last few decades floor systems have developed to include steel decking with concrete fill and finish. Nevertheless, all these offshoot developments fall under the **structural steel** framing system.

Whereas reinforced concrete was quite quickly taken into the building industry for use in foundations, peripheral cellar walls, and floor slabs supported by structural steel framing, it was some years before buildings with entirely reinforced concrete framing came into prevalent use. For many years now, however, reinforced concrete framing with masonry curtain walls has had an important part of the building industry.

As reinforced concrete was being perfected into higher strength systems with very predictable loading possibilities, precast concrete facade systems were developed to enclose structural steel and reinforced concrete frames. Then structural engineers developed precast concrete systems for entire structures, from framing system to enclosing walls.

As this text is being printed, new systems are being developed. They are new, but the basic "parents" of the building industry are the four types we have listed. If the student understands these four, he can instruct himself on the newer systems.

11–2   Wall-Bearing Construction

A wall-bearing building is exactly what the name implies. The walls support the entire structure. Of course, one could say that a frame house is a wall-bearing structure and it is. However, when we talk about wall-bearing structures we are usually talking about *masonry* wall-bearing structures. A masonry wall-bearing structure supports the inner structure (floors and roof) with the exterior masonry walls that enclose the building and, often, with interior load-bearing walls or partitions when the distance between two parallel exterior walls is too great for timbers (or steel members) to span.

A study of wall-bearing construction is important to every construction man. Currently, many one- and two-floor wall-bearing stores are being erected in shopping malls. Also, older wall-bearing structures require alterations, or new additions (not necessarily wall-bearing structures) are added to original units. Considering that many of our important public buildings and a considerable number of our older commercial structures are based on the wall-bearing system, it is entirely possible that at least half of our readers will be called upon to work with this system.

In wall-bearing construction the masonry bearing wall must support two things, the wall itself and the floor and roof framing. If the building is only one or two stories high, the masonry walls will probably be the same width (or thickness) for their entire height. However, as the height of the supporting masonry increases, the lower portions of this masonry must be increased in thickness so that this masonry will have the strength to support the masonry and floor loadings above. As the wall becomes higher and as there is less need for vertical support, this bearing wall can become thinner.

The logical place for a bearing wall to decrease in thickness is just below the joists so that these joists will have a **shelf** or bearing space to sit upon. If the floor joists meet the bearing wall at a level where there is no setback (i.e., no change of thickness of bearing wall), these joists will sit upon masonry; after the floor system is built, they will be built into the masonry. Figure 11–1 shows a typical bearing-wall situation with two exterior walls and one interior wall. Please note that both the exterior (peripheral) enclosing masonry walls and the interior (partition support) wall set back at vertical intervals. If the bearing wall between two separate buildings serves to support the floors of the two buildings, it is called a "party wall"; understandably, it must receive special care or protection when one of the buildings is being altered.

What we do not show in Figure 11–1 is what are termed **fire cuts** on the ends of joists embedded in masonry. Neither do we show the steel anchors tying the wood joists to the exterior masonry. The fire cuts are required by most building departments so that, in case of a fire, the joists will fall out of the masonry wall rather than pull the masonry down with them as they burn in the middle of their span. Fire cuts are pictured in Figure 11–2, which also shows steel anchors tied into masonry at the end of every fourth joist (or at least a minimum of 72 in.) and steel anchors, which are transverse to joists that must tie to at least three joists and be spaced at a maximum of 72 in. to satisfy most building codes. These anchors are required to tie the building together. However, in cases where a fire damaged the joists in one area to the point that they failed, the adjacent (nonanchored) joists would pull the anchored joists from the system and save the bearing walls.

When steel-rolled structural shapes became available, the spacing between bearing walls could be increased, and, if steel cross framing were

Figure 11–1. Typical wall-bearing system.

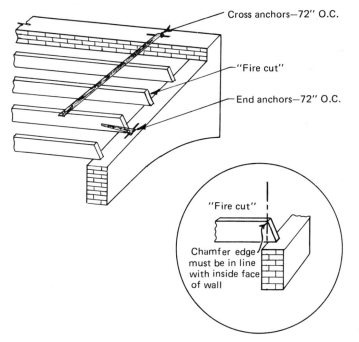

Cross anchors—72" O.C.

"Fire cut"

End anchors—72" O.C.

"Fire cut"

Chamfer edge
must be in line
with inside face
of wall

Figure 11–2.

used, considerable masonry supports could be omitted. However, as these steel joists and purlins imposed greater (per square inch) loads upon the masonry, steel base plates were required to spread the end load of the girders, just as footings are spread to lessen the per square foot loading onto the soil. Figure 11–3 shows how wooden joists can be supplanted by a steel

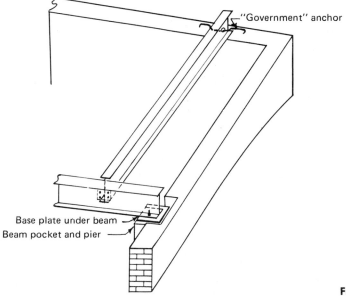

"Government" anchor

Base plate under beam
Beam pocket and pier

Figure 11–3.

girder-beam system when design requires. However, just as masonry–wood buildings must be tied together with anchors, steel supports must also be anchored to masonry to provide this homogeneity. The anchors holding the steel beams into the masonry at the ends are often called **Government Anchors,** a term that comes from their origin.

Figure 11–3 does not show an entire assembly for a floor's framing because such detail would cloud the important support framing and its anchorage into the exterior masonry. One must understand, also, that after the framing shown in Figures 11–2 and 11–3 is placed, along with masonry anchors and the flooring placed upon these supports, the masonry walls are continued upward to the next floor level. One advantage and economy of the wall-bearing system is that, because bricklayers can work from the floor just supported, interior scaffolding is only required for the completion of the top half of each lift.

11–3   Variations on Wall-Bearing Systems

Long before the steel column system was developed, the wooden column system came to be used in the wall-bearing type of design to allow larger and wider buildings, such as were used for factories.

Where now steel column sections are spliced together with splice plates and with bolts or welds and where the column has "clip angles" to support girders that join to the column, the wood column system used iron yokes to connect (or center) the wooden columns at splices; these yokes had flanges for the support of wooden girders that framed into these column splices. In turn, there were iron stirrups hanging onto the wooden girders to support purlins between girders (sheet-metal stirrups are still used to support floor systems in current wood construction). For an example of splicing of wooden columns, consider that a 20- by 20-in. wooden column is to be reduced to a 16- by 16-in. column. The 20-in. column will be cut (for length) to the bottom of the wooden girder line for the next lift. A cast iron collar slightly larger than 20 by 20 in. on the bottom and slightly larger than 16 by 16 in. at the top would sit on the top of this 20- by 20-in. wooden column. There would be centering devices on the overhanging flanges so that wooden girders would not slide sidewise. The 16- by 16-in. wooden column would be centered onto this cast-iron connection piece after the joists, purlins, and floor systems had been erected, and the masonry and wood structure would continue upward as each level was completed.

11–4   Precautions for Alteration Work

A superintendent for a building-wrecking company once stated that his profession was "reverse engineering." He explained that to alter or demolish a building without mishap he had to figure how it was **erected,** and then reverse this erection procedure in the dismantling procedure. No precautionary advice could be better stated.

Some older buildings were erected with most clever and sometimes devious methods. The superintendent for a razing or alteration job should not remove any structural member unless he is fully aware of its purpose and the method by which it was installed.

Another precaution that an alteration superintendent should take is to investigate whether the original structure still retains its originally designed strength. If it does not, the designer for the alteration work should be informed of the change so that he may order strengthening methods before new loadings or connections are imposed. In this regard, many row-house buildings of the past had **party walls** (see Fig. 11–1), which supported the floor systems of a building and adjacent structure and were jointly owned by the owners of the two adjacent structures. If an alteration plan calls for a change in a party wall, the superintendent must assure himself that the alterations he makes will not endanger the structure on the other side of the wall.

11–5   Current Wall-Bearing Systems

Whereas in cities, where land is expensive, office buildings and stores are housed in multistory buildings, the comparative lower cost of suburban land lends itself to one- and two-story stores and office buildings in large shopping malls. Because the land is less expensive, department stores can afford to have large, expansive display areas. And because these buildings are rarely higher than two stories, the internal support structure does not have to be intricate.

One of the most common wall-bearing systems in modern usage is a one-floor store building (often over 100 ft long) in which steel roof girders and steel bar joists are supported by steel columns on the interior and by masonry walls on the exterior (or periphery) of the building. In later years, because masonry costs are higher, these peripheral walls are concrete block; in former times they would have been brick. Brick is much stronger than most concrete block where there is a requirement to support vertical loadings. That is, the per square inch compressional strength is greater for brick. Therefore, in buildings where the peripheral walls are concrete block, brick piers (or pilasters) are built into the exterior, load-bearing walls where a girder is to rest.

This system is quite economical. As the only load that the columns (in a one-story building) must support is the roof system, these columns can be light. As the only load that the horizontal steel must support is at the most snow loading, the beams and bar joists can be light. With the exception of brick piers for the support of steel roof girders, the exterior masonry can be 12-in. concrete block or as designated by local code. Thus, generally, for a one-floor building this system is one of the most economical. But erection precautions are most important.

11–6   Precautions for Erection of Wall-Bearing
         Buildings

102

Types of
Superstructure
Construction

The usual procedure for the erection of any wall-bearing structure is to erect the exterior walls (and any interior load-bearing walls) to the first lift and then install the floor or roof system. In the case of large, low buildings, the steel floor or roof construction (including interior columns) takes but a few days with a crew of steel workers and a crane. For this reason the builder usually pours the column footings (complete with erection anchor bolts), the footings for peripheral walls, and then erects the peripheral walls and piers. After these walls are finished, the steel system is installed and, thereafter, the building is self-bracing.

However, until girders and bar joists span the building, exterior to exterior, there is no lateral bracing for the long, peripheral walls unless temporary wind bracing (see Fig. 11–4) is placed by the builder. This is a most commonly neglected precaution. Either through lack of knowledge or a misguided desire for cost saving, temporary wind bracing is omitted. Far too often entire walls, hundreds of feet long, are blown down by unexpected wind. The cost of bracing these walls is considerable. However, no bracing cost can compare with the cost of rebuilding the peripheral masonry. The intent of a good construction company is to erect a good building as cheaply as possible. However, the separate portions of the construction must stand up until all portions are erected and tied together. Be careful! Even though a superintendent's employer wishes to use every economy, the superintendent should protect his employer and himself by warning his employer when the desire for savings may endanger the structure. Always install temporary lateral bracing and keep it in place until the building is completed to the point that it can withstand these loadings.

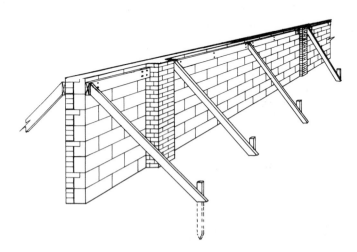

Figure 11–4.   Temporary wind bracing for bearing wall.

Chapter 12

Basic Structural
Steel-Framed Buildings

The first building to have a framing system that was completely structural steel was erected shortly after 1900. This building consisted of steel columns from footings to the roof and steel girders and beams supported by these columns. The steel girders and beams supported both the exterior masonry (now **curtain walls**) and the interior masonry partitions. Thus the then new structural system was one that was diametrically opposite to the former wall-bearing system in which the masonry supported the steel girders and beams. However, it had obvious advantages.

Primarily, the structural steel system allowed high-rise buildings that were not possible previously. Second, the steel framing could be fabricated in a shop where weather was of little problem and, as the steel was fabricated, it could be erected on the site in temperatures that were too low to allow the erection of unprotected masonry. Also, after the steel framework was erected, tarpaulins could be hung from this framing to provide wind protection and to envelope temporary heaters that could, then, be introduced into the construction system. With this wind and temperature protection, concrete and masonry could proceed in colder weather.

12-1   Accuracy of Steel Fabrication

If components of any building system are to fit together, each piece of shop-fabricated material must be accurately fabricated. Also, the completed system must end up with the correct final dimensions. Structural steel lends itself very well to these requirements. It can be cut to most accurate dimensions and, if greater accuracy in column length is required or if bearing transfer requirements demand, the steel can be milled on each end to insure great accuracy. Thus, insofar as the fabrication shop is concerned, there is no problem in producing a very accurate steel structure. However, even the best shop work can be voided if the erection procedures are not sufficiently accurate. Erection procedures fall under the jurisdiction of the Field Engineer and Project Superintendent. Thus, as this is a text for field construction, we shall dwell upon the methods of field erection and field engineering that will insure final accuracy.

12–2   Field Engineering to Insure Accuracy

104

Basic
Structural
Steel-Framed
Buildings

The accuracy of any structure depends on the accuracy of grade, or elevation, and line, the alignment of the structure in conjunction to its location to streets, property lines, and other important location requirements of the structure.

Primarily, grade is the first aspect of structural steel that a Field Engineer can check as structural steel is being erected. Columns of a structural steel building are set upon base plates or billet plates. Thus the accuracy (in elevation) of these plates will determine the accuracy of the elevation for any girder or beam in the *entire* building. Base plates and billet plates are set upon shims or, in some cases, there are elevation-setting bolts to support the plate from the concrete footing until the plate has been **grouted-in.** If the elevation of the plate is to be set by supporting bolts (usually three), the elevation of the plate can be set to $\frac{1}{1000}$ in. However, even if the plate is supported by shims, the elevation can still be relatively accurate. Steel erectors are supplied with shims of different thicknesses. Always included are shims as thin as ⅛ in. Thus, when the Field Engineer is checking the elevation of base plates soon after the erection of the first derrick columns, he should demand an elevation accurate to, at least, ⅛ inch.

Then, as soon as girders and beams for the first floor are set, the Field Engineer should set his level so that its telescope is no more than 5 ft above the first-floor structural steel and have his rodman run the rod over the entire floor-supporting steel. If the base or billet plates have been set for elevation to within ⅛ in., any horizontal steel of the first floor should be within ⅛ in. of the correct elevation. If the Field Engineer demands initial elevation accuracy to be within ⅛ in., he will have no problems in setting grades (4-ft marks) for finish floor at *any* level. "Start right, end right."

Linear accuracy is also most important. Primarily, lines for center line of footings and columns are set by the Field Engineer. Just before structural steel delivery is expected, he should set his transit onto column center lines and have his assistant mark the column center lines onto the footing. This is usually accomplished by painting a strip of yellow traffic paint onto the footing concrete and then, just as the paint hardens, drawing a black pencil line into the yellow paint. If the Field Engineer takes this precaution, he can be more readily assured that the column base plates will be set in the correct location and he will be able to check these locations more easily.

As columns and floor-framing systems are erected, they are held together with "field bolts" until the erection crew has set guy ropes (with turn-buckles) and has made the first tiers of columns (derricks) plumb. The erection crews endeavor to do this most accurately by hanging sufficiently heavy plumb bobs, and adjusting the turnbuckles on the temporary guy ropes until these plumb lines indicate that all columns are plumb in both directions. The Field Engineer checks this by setting his transit sufficiently distant from the building and slightly in front of the building line. After he has leveled his instrument, he sets the vertical cross hair onto the front corner of each column. He then raises his (plumb) telescope to the top of the column. If the column is not plumb (within tolerances), he will easily detect this and can advise the Steel Superintendent as to which columns need correction before finish bolting or finish welding may be done.

We have assumed that the Field Engineer has checked the location of outer building-line base plates with the center-line markings he placed onto the footings or the peripheral walls. However, he should *also* set his transit on the building line and check the outside of each column a few inches above the base plates. If this location is accurate and if the columns are checked for plumbness as the columns are erected, the entire steel structure (including interior columns) should be linearly correct. The steel can now be finish bolted or finish welded. Columns are usually fabricated in two-story lengths. Thus the first **derrick** (or erection system) will go from the cellar footings to the second-floor level, and the second derrick will go from the second to the fourth-floor level. Whereas the steel crew will have plumbing crews working continually, these crews tend to plumb four levels at a time for finish plumbing (i.e., two derricks). After each two derricks is approved by the Field Engineer and final connections are made, the crew moves up and starts plumbing the next two levels.

12–3   Responsibility for Correct Bolting or Welding

In the original structural steel system, columns were connected by splice plates and rivets, and girders and beams were connected by clip angles and rivets. The column splice plates were riveted onto one end of a column in the fabrication shop, as were the clip angles onto the beams. The columns and beams were erected and held together with field bolts until the Field Engineer approved line and grade, and then **field rivets** were driven. These field rivets were heated white hot and thrown (hot) to separate rivet crews who placed them into the holes and drove them tight while forming a head on one end. As the rivet cooled and shrinking (from temperature drop) was completed, the rivet became very tight. However, if the rivet was not hot enough or if it were improperly driven, a blow from a small rivet-testing hammer would show up this discrepancy. The tester would feel vibration.

In the days of field riveting, it was the responsibility of the Field Engineer (or his assistant) to walk the steel beams and check each rivet. When he found a loose rivet, he marked it with lumber crayon to be burned out and replaced by the rivet crew. Now, however, new connection systems and new requirements for connections have passed the responsibility for connections to testing laboratory personnel. The older system of riveting depended upon the rivet being tight and, also, on the hot metal of the rivet stem being expanded so that it completely filled the holes and allowed no "shear shift." In the newer bolting system, the bolting is accomplished with high-strength bolts that are tightened (or torqued up) to the point that their clamping action will not allow any shift. These bolts are tightened with air-powered impact socket wrenches.

If high-strength bolts are to be used, the impact wrenches must be checked and calibrated each day. The calibration is done by an inspector employed by a testing laboratory. The testing laboratory is retained by the design (structural) engineer or more usually by the owner. In addition to checking the workmen's air wrenches, this inspector spot-checks bolts after they have been finished with a calibrated torque wrench so that he can attest to the fact that the bolts are actually being tightened to the degree required by the American Society for Testing Materials.

If field connections are to be accomplished by welding, these welds will be made under the supervision of a welding expert supplied by a testing laboratory employed by the owner. He will watch welding procedures (especially the first "pass"), and will check each finished weld to see that it is correctly formed. If, for certain welds, the Design Engineer or the municipality requires that precise testing be made, these welds will be scrutinized by X ray, the "magnetic-particle system," or ultrasonic methods. The X-ray system is similar to medical X-ray investigation, but has the disadvantage that the method cannot be used when construction personnel or pedestrians are in the immediate neighborhood. If the project is in the suburbs or not in a crowded neighborhood, X rays can be made after working hours. If this is not possible, the magnetic-particle or the ultrasonic system could be utilized. In the magnetic-particle system the steel in the weld area is magnetized, and steel dust or particles are spread upon the surface of the welded area. The patterns in these particles, as formed by magnetic forces, may be "read" by an experienced operator who can judge the quality of the weld. In the ultrasonic system, sound waves are directed into the weld area and their reflections are "read" on a screen like an oscilloscope type of meter.

12–4   Floors for a Structural Steel-Framed Building

The floor system used on most structural steel-framed buildings is concrete or a combination of concrete on top of a steel bottom form. The most universally used system over the years has been the flat slab system wherein a concrete slab is formed over plywood forms so that the finished concrete slab rests on top of the steel beams. Where there is a desire for fireproofing of the steel beams, forms are formed around the beams (see Fig. 19-1a and b).

However, with the advent of underfloor electrical circuits and the rise in cost of wood forms, several steel-decking companies have come out with systems that not only provide the bottom form for the concrete slab, but provide strength to the floor system so that a thinner (or less strong) concrete slab may be used. In addition, these metal deck systems provide raceways for the underfloor wiring.

One of many steel decking–concrete floor systems is shown in Figure 12–1, a system of tubes or cells whose sections tend to provide considerable support to the floor and, also, are used as raceways for wiring. Other sys-

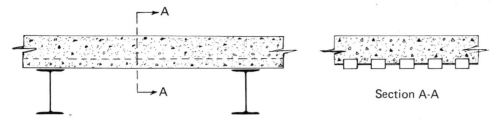

Figure 12–1.

tems employ a type of corrugated steel sheet for the bottom form (often with bars or mesh welded onto the corrugated steel so that the form system is bonded to the concrete) and may be used for extreme-fiber stresses, along with a separate raceway system for underfloor wires.

12–5   Facades for a Structural Steel-Framed Building

Whether the facade of a building is to go on a structural steel frame or onto a totally reinforced concrete building, the facade should start as soon as the framing structure of the building is advanced enough to allow this next operation to commence. In the steel-framed building, the next operation after steel is plumbed and connections are complete will be the installation of the concrete floor (or steel deck and concrete floor) system. In fact, in most codes, structural steel may not be higher than eight floors above the last concrete pour. However, once the pouring of concrete floors (or arches) is commenced, the facade should not be too many floors behind the pouring of floor systems.

The facade of a steel-framed building can be of many systems because, primarily, the strength of the structure is provided by the structural steel. However, the facade system can cause additional loadings on the steel framing system. To explain, recall the original facade system, for which, after the steel framing and the concrete floors were erected, a masonry facade that used normal-sized sash was built. Because the sash did not take up the entire bay as do glass-facade buildings or buildings with continuous sash, there was room at the top of each column for **knee bracing** (i.e., bracing for horizontal wind or sway bracing; see Fig. 12–2). This knee bracing would not show because it could be between sash at the columns. However, when designs provided for continuous sash or facades of glass, this knee bracing had to be replaced with other sway bracing, which usually led to X-bracing in the core of the building.

Regardless of the facade system, the facade must be tied to the columns and the floor systems. If the facade is a conventional built-in-place masonry system, it will be tied to the columns with anchors. If it is a spandrel-window system, there will be a need for anchorage to the spandrel beams or

107

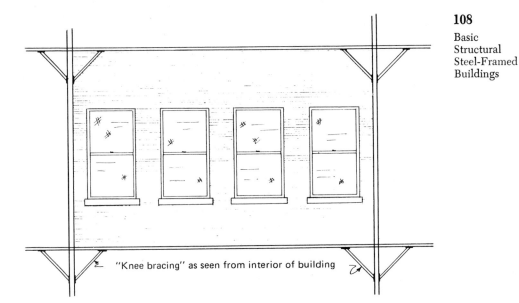

"Knee bracing" as seen from interior of building

Figure 12–2.

floor slabs in addition to the column anchorage. If the facade is to be glass, precast concrete, or any other factory prefabricated system, there will be special anchorage requirements that the superintendent must make allowance for before concrete floors are poured.

Prefabricated facades are a great help to the progress of a project. Primarily, they can be fabricated under controlled conditions and may be inspected by the designer at the fabricating plant so that, when the project needs the facade sections, they will have been inspected, approved, and ready for erection.

Chapter 13

Reinforced Concrete Framing: Poured in Place

Poured-in-place concrete has a great advantage for the contractor. Primarily, he can set up his own schedule and can pour as much of the building, per pour, as the size of his field organization will allow. Second, his materials (i.e., concrete and reinforcing steel) are quickly available. Reinforced concrete buildings were limited in height for many years. However, now that concrete contractors are more knowledgeable and more reliable, many municipalities are allowing concrete structures to be designed with less safety factor. As an explanation, consider the city that once listed in its building code

1. Class A concrete.
2. Class B concrete.
3. Controlled concrete.

Controlled concrete was concrete that was **batched** under the supervision of a laboratory inspector, had on-site test cylinders made by a laboratory inspector, which, at intervals of 7 and 28 days, were tested for strength by the laboratory. Class A concrete was not a better mix than class B concrete. Class A concrete, as specified by that building code, was designed for class A contractors who could be trusted to do a good job and not add too much water to the mix. The class B design mix was to be used by class B contractors and had more cement per yard so that these "class B" contractors could pour it like soup and still come up with fairly reputable concrete.

In the days when class A and class B concrete were specified, building codes demanded high safety factors. Thus it took more concrete (i.e., heavier structural members) to achieve required strength. This made the building heavier and, in turn, the heavier building limited the height. Lately, however, three things have happened:

1. Chemical admixtures have been developed that aid concrete to flow more easily at lower water–cement ratios, so there is less need to add (strength-losing) water to help place the concrete.

2. Concrete-mix designers have proved that it is important to have good blending of concrete aggregates (the sand, small stone, and larger stone), and mix designers are making much progress in this endeavor.

3. Concrete users in general have learned the importance of pouring concrete as the particular mix was designed, and have become more dependable in doing this and in protecting the concrete after it has been poured.

110

Reinforced
Concrete
Framing:
Poured in
Place

In addition, because of the greater dependability of concrete mixes and the contractors that use them, and because of pressure from design engineers, building codes are allowing lower safety factors on the condition that the mixing and placing of concrete be controlled, and that test cylinders be taken for all concrete. Thus lesser amounts or weights of good quality concrete are needed to fulfill a design.

The result of these advances has made concrete a more competitive material. However, the contractor must both *respect* his concrete and the manner in which he pours it. In fact, a contractor who specializes in pouring fine concrete has an affection for the material because he knows he can depend on it for a good project.

13–1   Field Engineering to Insure Accuracy of Line and Grade

The field engineering for a concrete building is little different than that used on a steel-framed building. Center lines are placed upon footings so that the carpenters can accurately set the column forms. After the forms are set, the Field Engineer must check them for location and plumbness by setting his transit on an offset line and taking readings (distances) on the column forms, at top and bottom of the column form.

There are many procedures in pouring a concrete building. Columns are usually formed and braced (with reinforcing placed, of course) and slab forms erected before columns are poured. In many cases the concrete beams and slab will have steel reinforcing bars placed and the whole thing (columns, beams, and slab) will be poured at the same time. Of course, if this system is chosen, concrete column forms should be filled to the level of horizontal concrete (i.e., beam bottom or bottom of flat slab) and allowed to set for at least 1 hour before the beam or slab concrete is poured, so that any concrete shrinkage in column concrete will have taken place before the horizontal concrete (the column's eventual load) is poured on top of it.

If it is at all possible, the author prefers to pour all columns to beam-bottom grade first, sweep out the beam and slab forms, place the steel reinforcing for the beams and slab, and *then* pour the beams and slab as a second pour. Many standard and city codes require this two-pour system. If the two-pour system is used, the concrete columns (after concrete is hard) will help stabilize the alignment of the building, and easier and more accurate pouring of the beams and slab will be possible. To explain the latter statement, if all the reinforcing for the beams and slab is in place when the columns are poured, there is a great chance that concrete intended for the columns will spill into the horizontal forms before one is ready to pour this horizontal concrete. If the column concrete has not had time for full shrinkage, any spillage should be removed. This secondary cleanup seldom happens. The building ends up with "cold joints" and poorer concrete.

After the first slab has been allowed to set for 12 to 24 hours, the Field Engineer should be shooting an axis line (usually a line 2 ft off a column center line or halfway between center lines) onto the new concrete deck in both directions (i.e., at 90° to each other). To insure that these lines are not obliterated or lost, he will have his rodman or assistant drive "crete" nails into rough-pour slab surfaces (i.e., raked for future topping) or scratch lines (at intervals) into a troweled surface. In this way the carpenters can snap chalk lines between these references and have accurate lines with which to set the next lift of columns.

Grade for concrete structures is no different than grade for steel-framed structures, except that the Field Engineer has to set grades constantly. With the steel frame (see Section 12–2) it was sufficient to check the elevations of all base plates, and later to check the elevation of all first-level steel beams. However, with reinforced concrete structures, the accuracy of the building depends totally on the work of the Field Engineers and the carpenters. As soon as a column is formed, the Field Engineer will set a 4-ft grade (i.e., a mark 4 ft above design finish grade). This will give the carpenters exact information so that the column form can be cut for beam bottoms and slab bottom. These 4-ft marks on the columns will also be used by the carpenters to stretch string lines to set the elevation of the slab forms.

After the column forms have been plumbed and checked by the Field Engineer, he will set his surveyor's level under the forms for the slab, and will read an inverted rod to check the level of the beams and slab forms. If beam forms are large (i.e., there will be a heavy concrete load), or if a beam is to be a long one, the center of the beam should be set slightly higher, or **cambered,** so that, if there is any settlement, there will be no sag in the beam. A long beam looks better with a slight camber than a sag. Actually, if the beam ends up dead level it may seem (because of optical illusion) to have a sag. As soon as the Field Engineer has checked the beam and slab forms, he will set up his level on the top of the forms in a solid position on a previously poured section of the same slab or over a previously poured column, and set the screeds for **top of concrete.** Then, shortly after this level is poured, he will start all over again!

13–2   Inserts in Concrete Pours

When the structural frame of a building is steel, anchorage for facade sections, relieving angles, or exterior architectural features can always be added if they have not, already, been welded to the frame. However, with a structural concrete frame system, anchors and inserts **must** be installed prior to pouring. It is the responsibility of the Field Engineer and carpenter foreman to be sure that all anchors and inserts are in place. However, remember that we are discussing anchors and inserts for structural and architectural features. In addition to anchors for these purposes, there will be slotted anchorage for elevator rails, which will be supplied by the elevator contractor and installed by the carpenters. There are other sleeves and inserts that will be installed to handle the mechanical/electrical trade requirements. The setting of these sleeves and inserts is the responsibility of the trade involved. However, the Project Superintendent and the Field Engineer will wish to be sure that these trades have completed this work prior to pouring.

112
Reinforced
Concrete
Framing:
Poured in
Place

13–3   Facades for a Structural Concrete-Framed Building

There are no limitations as to the type of facade that may be built onto a reinforced concrete frame. Facades may range from the original masonry systems to spandrel plus sash, all glass, to the precast concrete facades that are used on the structural steel frames. However, as noted in Section 13–2, anchorage inserts must be installed for these installations. In the case of masonry facades, there must be preparations for relieving angles, and there must be slots in the columns for "dovetail" anchors in the masonry. In the case of precast concrete facade sections, there may be a need for small bearing plates to be cast into the floor over the spandrel beams. However, other than the preparation and inserts required, there are few limitations on the type of facade that may be installed on a reinforced concrete building.

13–4   Advantages of a Reinforced Concrete Building System

Whereas the structural steel framing system has the advantage of prefabrication and erection in colder weather, the reinforced concrete framing system has the advantage that framing can start as soon as reinforcing steel is available. Thus, when a building is to be constructed from plans that cannot reach the contractors and shop-drawing people much before start of construction, or in the case of "out-of-phase" construction for which excavation and foundations for a structure are being done long before final designs for the entire project are completed (and thus the structure should start soon after), the reinforced concrete system has decided advantages. Shop drawings for steel reinforcing can be made for the lower portions of the building, and can be prepared and approved lift by lift as the building rises. Except in situations where reinforcing steel is in short supply, a reinforced concrete structure has the advantage that it can commence immediately.

113

Disadvantages
of a
Reinforced
Concrete
Building
System

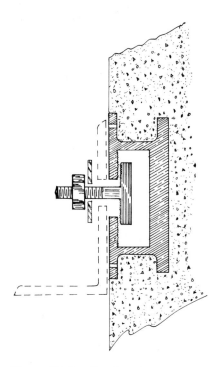

Figure 13–1. Cast iron insert to support a relieving angle in structural concrete.

13–5 Disadvantages of a Reinforced Concrete Building System

The advantages of other building systems sometimes point to the disadvantages of the concrete frame system. One that occurs in every phase of concrete pouring, whether for the total concrete frame or the concrete slab on a steel frame, is that the poured concrete must be protected from the weather. When pouring a floor on a structural steel frame, the structural steel around and above the floor being poured may be used to support tarpaulins. There is no such upper framing in the reinforced concrete system. True, tarpaulins can be hung around the column forms and the periphery of the floor (and/or columns to be poured) so that temporary heat under the pour can be contained. However, there can be no protection from the elements for the concrete above until it is poured and has set up enough to support protective tarpaulins (or craft paper to protect new concrete from fast drying). Thus, with a steel frame, the contractor might chance a pour when there was to be a possibility of rain because he could cover portions of the pouring by hanging tarpaulins over the structural steel of the next floor. With the reinforced concrete frame, the contractor may have to watch weather conditions more closely because it is more difficult for him to protect top surfaces.

Chapter 14

Completely Precast Buildings

Buildings which are constructed of completely precast concrete units are not new in the world's building industry, although they are somewhat new in the Western world's building industry. In Russia precast buildings were being erected before 1940. In the Russian buildings (originally apartment buildings), everything was precast, including the footings (which were set onto a thin layer of job-poured concrete to transmit the load of the footing onto the earth or rock below). In the original Russian system, gantry cranes were set up on construction sites prior to construction, and then every piece of construction material was lifted from trucks and set into place. Bearing walls (when they were interior partitions) had plumbing piping cast within. The ends of these pipes were tapered so that, as the upper partition was lowered into place, the pipes meshed and the taper assured a tight joint. At that time, this system could have been accomplished only in Russia, which did not have union problems. In Russia the workers accomplished this type of coordinated, prefabricated construction for the common good. In the Western world, trade unions would have insisted on installing all piping and electrical conduit on the site.

A prerequisite of any precast system was fulfilled by the original Russian system. Primarily, a precast system will not be economical unless the same forms or molds can be used many times. To explain, the metal or fiber-glass molds for precasting concrete are very expensive; they must be utilized often before their cost is absorbed. This requires the acceptance of many buildings that are completely similar, a condition that many American architects deplore. In Russia, this was no problem. The workers were happy to have the lower-cost housing, regardless of repetition. They cooperated with the system.

Union problems in many countries, especially in the United States, deterred introduction of precast systems in which utilities were cast into the concrete until the great need for less expensive housing brought builders and unions together. Then the precast systems began to flourish, first in Europe, then in England, and then in the United States. All these systems require some indulgence from union members in that the "shop" union must

Figure 14–1. Precast concrete building in the erection process. (*Courtesy of Strescon Industries*)

install conduit, steel reinforcing, and items that would, normally, be installed by other unions in the field. However, now, when low-cost housing is necessary and when everybody needs work, these systems are progressing throughout the world.

Another prerequisite for prefabrication economy is that the prefabrication plant be near the sites where buildings are required. It does not make sense to spend the money saved by prefabrication for trucking the components many miles to the construction site. Many prefabrication plants do in fact fall near to areas that require many units. Therefore, there is a distinct saving in trucking, besides the advantage that the building is being precast in a controlled environment while the weather at the building site may not be suitable for poured-in-place concrete.

There is another, most distinct advantage to prefabricated concrete components. In any system the finished product is subject to the approval of the designer. In the finished product (in this case, perhaps, exposed-aggregate facade panels) the units may be inspected by the architect as they are cast. Thus, when erection of the units is required, the units will have been accepted by the architect and may be installed in the building. Were the units to be cast in place on the project, a rejection could cause considerable problems, which, if they could not be solved by "cosmetic" remedies, would require replacement. Knowing in advance that components have been accepted is a valuable asset. If the prefabricated units have previously been accepted, the building can be erected immediately after footings have been poured and have cured enough to achieve enough bearing value. Because the prefabricated units have been cured in warm fabrication plants (usually by steam-curing methods), the bearing units can be utilized for their support ability immediately after they reach the project.

Precast concrete buildings come in many diversified systems. Generally, however, there are bearing walls, shear walls, and/or bearing-providing partition members that support large flooring concrete plank units. As soon as footings are ready to support the load for which they were designed, the bearing units can be placed upon them. These units will be leveled with shims, bolted down, and held plumb (see Fig. 14–2) with temporary braces bolted to the footings or floor planks.

Figure 14–2. *(Courtesy of Strescon Industries)*

Thereafter, the concrete floor planks will be set (with bearing-wall bolts going through for upper bearing wall connections), and then the next level or "lift" of bearing walls or partitions will be set, shimmed to level, and braced. If the ambient temperature allows, these units will be grouted with nonshrink, "dry-pack" cement mix. If temperature conditions do not allow, several lifts may be erected before grouting is required. Then temporary heat can be brought into the structure so that the grouting may be placed and protected until cured, as the structure above proceeds.

As soon as several lifts of the precast building are erected, bolted down, and grouted, the mechanical/electrical trades can commence with their installations and follow the rise of the building. If the progress schedule has been carefully set forth, the mechanical/electrical contractors will have secured their long-lead equipment, and this can be set as soon as the space for which it was intended has been erected.

117

Proper
Scheduling
for a
Precast
Structure

14–1   Proper Scheduling for a Precast Structure

As we have previously indicated, advantages of the precast system are that the units may be fabricated in a factory atmosphere where outside temperature is not a problem, and that these units can be inspected and approved long before they arrive at the site. However, if the builder or construction manager is to fully utilize these advantages, he must carefully schedule the other operations of the project.

For example, in the structural steel framing system, the steel contract must be awarded long before the excavations and foundation work are commenced so that, as foundations are ready, the structural steel supplier will have ample time for the preparation and approval of shop drawings and the fabrication of the structural steel. After the structural steel frame arrives on the project, there will be considerable time before the mechanical/electrical trades are required for anything but the installation of sleeves and inserts.

However, in the precast concrete system, the floor planks will have holes for pipe or duct chases already cast in, and any large pipe holes will also be cast in the plant. Holes for smaller piping will be cored out in accordance with previous allowances and factory preparation of concrete planks. Thus installation of mechanical/electrical systems may start soon after precast units are erected. This erection ability places more responsibility on the mechanical/electrical contractors for ordering their long-lead equipment so that it will be available sooner than would be required by other structural systems. The same responsibility falls upon the suppliers and installers of such items as sash, stoves, plumbing fixtures, elevators, hardware, doors, and furnishings. For example, the author has been involved in two recent projects where high-rise precast structures were available to tenants four months after foundations were completed.

14–2   Field Engineering Control for a Precast
Structure

As is true of any type of structure, engineering starts at the bottom and progresses upward. If the precast structure is to be accurately located and plumb, the Field Engineer must be sure that the footings and the anchor bolts for the precast bearing units are in the correct location. Thereafter he must check the elevation of the bearing walls and shear walls as they are set onto shims to be sure that the *initial* elevations are correct. Then, as the structure rises, he must check the erectors' work to be sure that the panels are plumb and that new bearing panels are at the correct elevation so that the floor systems will be level. The erectors of bearing members will (probably) use 8-ft plumb boards, usually made of aluminum "I" bars that are very much like the bricklayer's 4-ft plumb level. The walls will be plumb and will usually be checked by the erectors to insure that they are exactly above lower bearing units. However, the Field Engineer should check this with his transit, and he should check levels at each floor as the erection proceeds. Engineering checks are important in precast systems especially, because current designs usually call for long lines of sash. The installation of sash in this manner requires level and plumb apertures.

14–3   Other Considerations with the Precast
System

Inasmuch as the precast building is made up of many units that must be connected in the field, shop preparation must be made for the accuracy and adequacy of these field connections. When the connections are interior, the grouting must be well done and well cured. When the connections are exterior, there must be allowance (in the grouting or in the jointing) for the application of sealants between the panels and between the panels and sash units to insure that moisture cannot enter. Also, at roof elevations, **reglets**[1] and provision for roofing and "blocking" must be cast into upper members. Too many of the early prefabs had leaks between facade units that were difficult to correct. More investigation during the design portion of the buildings would have precluded this very unhappy deficiency. As the building is erected, care in the installation of the sealant backer rope, the precleaning of concrete and sash surfaces to receive sealant, and the care in which sash, air-conditioning sleeves, and waterproofing systems are installed is most important. Chapter 21 will explain backer ropes and sealant systems.

14–4   Recapitulation of Precast Systems

Everything costs money. The fabrication and erection of a building costs money and the time of erection costs money. Prefabrication of the major portion of a building is a costly process. However, if the fabrication plant is

---

[1] A reglet is a preformed slot inserted to receive roofing and/or flashing.

relatively near the erection site and if weather conditions (e.g., cold weather) would slow a project, there are considerable advantages afforded by the precast systems. The "break-even" line depends primarily on delivery costs, weather conditions, and the money that can be saved by quicker use of the finished building. There is, in addition, a great advantage in knowing that all exterior finishes are approved long before they are required. For example, when concrete is cast in place, a rejection for reasons of surface appearance could cause considerable extra costs for the builder and even the owner.

Assuredly, then, there are many advantages to the precast building system if local conditions and supply allow. However, if extra costs of prefabrication will overbalance final economies, every advantage of the precast system must be used. Hopefully, the same general design can be used for a number of buildings so that the molds or forms can be reused to the point that their initial extra cost is absorbed. And if we are paying extra money for prefabrication, we should expedite footing installation so that erection of the first building unit may commence as soon as precast members are ready for delivery. Then, too, every advantage of careful scheduling and early ordering of long-lead equipment must be exercised.

As soon as the structure allows, sash and glazing must be installed so that gypsum board partitions, closets, and other architectural appurtenances (which require warmer temperatures) may be installed and spackled. Then stoves, plumbing fixtures, and other equipment may be installed.

If there is a universal bylaw or rule for the building industry, it is "get it done as soon as possible and as fast as possible." Prefabrication costs extra money in many cases, but the extra money spent in the first, prefabrication, portion of the project can be recaptured and additional money can be saved in the erection and finishing portions of the project with a well-expedited schedule.

Chapter 15

Miscellaneous Iron and Architectural Metals

Miscellaneous iron includes all the iron and steel members that are not supplied by the structural steel contractor on a structural steel-framed building, and includes **relieving angles**[1] for the support of masonry facades of structural concrete-framed buildings. One of the most important items supplied by a miscellaneous iron contractor is metal stairs. Understandably, the stringers and pans for stairs in a structural steel-framed building must be made from steel. The stairs for structural concrete-framed buildings are often concrete formed and poured at the site. The stairs for totally precast concrete buildings are usually precast units set into the building in conjunction with the erection of the precast structure. Nevertheless, steel stairs are often used in poured or precast concrete structures. Also included in the list of miscellaneous iron items are castings for the nosings of concrete stairs and for the concrete that is poured into the pans of steel stairs, lintels to support masonry, special tie rods such as those used to tie masonry facades to steel columns, pipe railings for stairs and platforms (or pits), trench covers, channel-iron door bucks, areaway gratings, and hundreds of items that can be made from castings, basic steel members, and heavy sheets of steel.

Architectural metals usually includes items fabricated from stainless steel (not including kitchen-equipment items, which is a specialty in itself) and building specialties made from nonferrous metals such as brass, bronze, and anodized aluminum. The Architectural Metals division usually includes the exterior-finish metals, such as storefronts, exposed stairs, ornamental doors and frames, revolving doors, exterior railings for entrance walkways, exterior air-conditioning louvers, flagpoles, and scores of items that are made from the fancier metals. Metal spandrels and window units are considered to be architectural metals, but are now separately specified and are usually fabricated and set by a company that specializes in these items.

One might think of miscellaneous iron as covering the *supporting* items of the building, which are needed earlier, and architectural metals as cover-

[1] Relieving angles are angles bolted to the structure and supported from the structure to support *another* portion of the building (usually masonry).

ing *finish* items, which are installed later. Ordering sequences that are important in both divisions, and certain precautions that should be observed will be discussed in the following.

15–1 Miscellaneous Iron: Sequence of Delivery

The management of a firm that specializes in miscellaneous iron knows which items come first, generally. However, if the General Contractor plans to call for certain construction first and that portion of the construction involves a miscellaneous iron item, he should alert his subcontractor. As soon as a miscellaneous iron subcontract is awarded and signed, the subcontractor should receive a listing of the sequence in which the General Contractor will require items. This listing will determine the order in which the subcontractor's shop drawings will be drawn and submitted.

For instance, if the stairs for the fire stairways are to be steel, the subcontract for Miscellaneous Iron should be bought soon after Structural Steel is under way. Thereafter, the subcontractor should be pushed to make and submit the stair drawings, and the architect should be pushed to review these drawings. As soon as the stairway shop drawings are approved to such a degree that fabrication can be commenced, the General Contractor should set up a delivery schedule with his Miscellaneous Iron subcontractor and follow the fabrication of stairs to the point that he may send an expeditor to that plant to be sure that the stairs will arrive on time. The timely delivery of stairs is important to the project. Workmen will climb ladders to a *limited* number of floors. Thus steel stairs should not be more than four levels behind the erection of structural steel.

There may be certain items of miscellaneous iron that should be on the project prior to stairways. Angle frames for pits and areaways that must be set prior to pouring of foundation concrete, corner guards that must be poured with lower concrete, and special frames that must go into basement walls fall into this category. If these frames are to be exposed to the exterior, they must be hot dipped galvanized. This galvanizing takes extra time and thus the shop drawings for these items should be expedited.

If stairs are to be poured-in-place concrete, stair-nosing castings will be required as soon as *floor construction* has passed the third floor. These castings are made to order, and are, therefore, a long-lead item. The shop drawings for railings for structural concrete stairs will usually show the nosings and will indicate the sleeves that must be cast into the concrete of the stairs as the nosings. Later, rail supports will be inserted in these sleeves. Thus, if the shop drawings for the railings and nosings are approved in time for nosings to be cast and delivered to the project, there will be no trouble in having rail sleeves on the project in time.

Next in delivery importance are frames, inserts, and anchors that must be set onto and poured with structural concrete floors; if the frame of the building is structural concrete, inserts for relieving angles will be in this grouping. Thus relieving angles and hung lintels (which would be supplied along with the structural steel on a steel-framed building) are next on the "want list" so that a masonry facade may be erected as soon as possible. If the facade of a concrete-framed building is to be one of the "skins," the anchorage inserts are all that will be required.

As soon as exterior or interior masonry is started, there are bound to be certain openings that require channel-iron door bucks. Thus these are needed next, on the "want parade," quickly followed by flagpole sleeves and cast-iron wheel guards for garage-type doors. Every project has special needs; understandably, there are special items in the miscellaneous iron division that must be supplied, and the sequence of delivery will depend on the special needs of each project.

15–2   Preparation and Protection of Miscellaneous Iron

Almost every good specification requires that the miscellaneous iron subcontractor remove any mill scale, rust, or blemishes and give the iron a coating of red oxide or zinc chromate primer paint. However, year after year, as costs rise, this is one area where the fabricator skimps. Zinc chromate primer is a good primer if it is applied thickly enough. However, it seems easier to skimp (by thinning) with a zinc chromate primer than with a red oxide primer paint. Most specifications require that the painting contractor wire brush all rusted portions of ferrous metals and "spot prime" them before proceeding with finish painting. However, repairing the primer on a rusted member is expensive and, unless the remedial priming is excellent, it does not do as good a protection job as a priming that was done well **originally.** Thus it behooves the Construction Superintendent to make his miscellaneous iron subcontractor fully aware that he will expect a good priming job and will reject any iron with thin paint or paint applied over mill scale. Another caution: if iron comes onto the project with thin paint, voids in paint (which we call "holidays"), or rusted areas, these problems should be remedied **immediately.** The longer they stay on the job the rustier they will become. In that regard, the Project Superintendent should advise his miscellaneous iron subcontractor that an accomplished painter should do this remedial wire brushing and repainting as soon as erection is completed. Brushed-on paint should be insisted upon. Do not allow remedial work to be done with aerosol cans of primer paint. Spray cans do not deliver paint in a thick enough consistency. Do not wait until the painting subcontractor appears on the project to have poorly primed iron painted by him and backcharged to the ironworker. Too much time will elapse.

If miscellaneous iron is to be covered with masonry mortar, this iron should receive a coating of black asphaltic paint over the prime paint before the iron is embedded in mortar. If the lintels over an *interior* door buck rust, not too much harm will occur. However, if the lime content of the mortar or general moisture rusts relieving angles or lintels over exterior openings, this rust may eventually seep out to the surface of the facade, and there can be no remedy but to remove the masonry and repair the rust situation. This can be very costly. Be *sure* that the iron is correctly painted in the first place!

123

Architectural
Metals:
Shop
Drawings
and Delivery

15–3 Architectural Metals: Shop Drawings and Delivery

Architectural metals are used as the finishes or **veneers** on the interior and exterior of the building. Because the finished products must fit or match these items, great care must be put into the shop drawings, which are usually drawn in larger scale than other shop drawings and show considerably more detail.

In discussing the subject of shop drawings we must note that the subdivisions in which certain architects place materials that must be used in conjunction with the architectural metal are not always correct. For instance, there are times when the Architectural Metals specifications require that this subcontractor take a channel-iron door buck provided under the Miscellaneous Iron subdivision and a hollow-metal door provided under the Hollow-Metal subdivision and "clad" both of them with stainless steel. A knowledgeable General Contractor or Construction Management organization will relieve the miscellaneous iron contractor and the hollow-metal contractor from the requirements of manufacturing and supplying the basic materials, and will place this responsibility onto the architectural metals contractor. This will preclude any "division of responsibility" and will make the architectural metals contractor responsible for the whole opening, including its stainless steel finish. This contractor may *actually* buy the channel-iron buck from the project's miscellaneous iron contractor and the hollow-metal door from the project's hollow-metal contractor. Nevertheless, the architectural metals contractor will be responsible for coordinating all three sets of shop drawings and seeing that the hardware (usually put into the architectural metals section for special openings) fits the buck and the door *after* cladding; thus he is responsible to see that an entire opening operates as it should and with the required clearances.

Architectural metal includes brass and bronze. In the last few years, however, actual brass and bronze finishes have been supplanted by less expensive anodized aluminum, which is anodized to look like certain brasses or bronzes. The application of these anodized finishes or treatments cannot be made until after the aluminum item (or portion of an item, in the case of large items) is fabricated. Also, certain special aluminum alloys are required for certain anodized colors. Large aluminum smelters do not batch up a few ingots of certain alloys on a moment's notice. Thus anodized aluminum in certain finishes requires months and months between the time that the base material is ordered from the smelter and the final member is extruded, fabricated, anodized, and ready for installation. Even if the color or finish

required uses a more readily obtainable alloy, considerable time is required
between buying the subcontract, preparation and approval of shop drawings, and delivery of the project. Thus the watchword for architectural
metals is buy out early and push the subcontractor. At the same time, the
architect must be pushed so that shop drawings are returned as soon as
possible.

15–4   Protection of Architectural Metals

A finished piece of architectural metal represents considerable time of a
fabricating artisan and a considerable amount of money. The cost of heavy
paper, masking tape, board protection, and even barricades along with the
cost of the carpenter and helper who install protection is nothing alongside
the cost of the metal needing protection. Then, too, one must consider the
long time from fabrication shop to the installation stage; replacement of the
item *after* installaton would be a severe setback to the project's completion
schedule. Therefore,

1. Be sure that the material is delivered to the project with sufficient
    protective coating or crating, and that this protection remains on the
    item until it can be stored in a safe place.

2. As soon as the material is uncovered or uncrated, have it inspected
    to be sure that the item is in good shape in case it is scarred and must
    be replaced. If replacement is necessary, do not allow an architec-
    tural metals subcontractor to talk you into accepting a blemished
    item because replacement will retard the schedule. That's *his* prob-
    lem. Get the material replaced, especially if it's at the lower floors or
    close to view.

3. As soon as a section of architectural metal is finally installed, have it
    covered with paper, cardboard, building paperboard, lumber, or
    whatever is required to make sure that the finish is unblemished
    until after the project is finished or ready for use in that area.

15–5   Color of Architectural Metals

One would think that stainless steel has only one color. But unless pieces
of stainless steel have the same finish-modulus (i.e., degree of smoothness),
and the finishing lines are in the same direction, adjacent pieces of stainless
steel will appear to have different colors.

It is understandable that brasses and bronzes must be of the same alloy
throughout in order that colors of adjacent pieces of finished materials will
match. But when we come to the anodized aluminum surfaces, we have big
problems unless the fabricator is careful. In the first place, no anodizing
shop will guarantee to have every piece of anodized aluminum the *exact*

same color. Instead the anodizor will guarantee to have all pieces in the same *range* of color. That is, he will give the architect two pieces of anodized aluminum and guarantee that all pieces of an order will be of a color that will fall between the two extremes. The architect will accept such a range. However, when the fabricator of a storefront or a column cover receives a number of identical anodized extrusions that he must piece together and cut to fit top, bottom, and side conditions, he should choose between the extrusions before cutting them to be sure that an extrusion from the darkest end of the range is not placed adjacent to an extrusion from the lightest end of the range. Even though the architect has approved a color range, he has every right to reject a frame or a column cover that has the lightest light extrusion next to the darkest dark extrusion.

Thus the Project Superintendent would be wise to watch the material of the architectural metals subcontractor as it comes onto the project and reject any material or fabricated section if he does not feel that it will be acceptable to the Architect or owners. If there is a borderline case, he would be wise to seek the Architect's approval of the section before it is built in.

15–6   Recapitulation

A building could not be erected unless there were lintels, relieving angles, and other supporting iron, no more than an automobile could be driven with three tires. The finished interior and the finished exterior of a building would not be acceptable, even if the partition facing or facade facing were of the most beautiful marble, if a minor amount of architectural metal in the same areas is not acceptably finished and installed, any more than an automobile will look acceptable with dirty windows.

The two materials, like the fourth tire or the glass of the automobile, seem to be minor when one considers the entire project. But no chain is effective without all the links being of equal value or strength. Thus, although the Miscellaneous Iron division and the Architectural Metals division may seem small, they are very important to the successful completion of the total project. These trades should be "bought out" early. Their shop drawings should be produced quickly and approved with expedience. And the final materials must be carefully fabricated and carefully protected.

Chapter 16

Concrete: A Major Construction Material

We have discussed foundations, including piles and caissons, which require concrete, we have discussed steel-frame structures with concrete decks, and we have discussed all-concrete structures. We have learned that concrete is versatile and that it figures in almost every type of construction. Now let us discuss the material itself and the different ways it may be monitored and manipulated so that it will serve the construction industry best. It is important that we fully understand the material if we are to use it sensibly and correctly.

16–1   Origin of Cements and Concrete

Concrete has been used in construction even before the time of Julius Caesar. However, the material used in those days was not a true hydraulic material such as we use today, but rather a mixture of hydrated lime and volcanic ash. This product is known today as Puzzolan cement. A product closer to what we use today (i.e., one that hardens under water) was developed in 1756 by the British engineer, John Smeaton, for the construction of the Eddystone Lighthouse. In 1796, another Englishman, James Parker, made the first natural cement by calcinating and grinding an argillaceous limestone. In 1824 a British engineer, Joseph Aspdin, patented **portland cement,** a much superior product. This cement was made by careful grinding of limestone and hardened clay, burning of a specified mixture, and then regrinding the product. This was named portland cement because it closely resembled the building stone quarried from the island of Portland.

In the United States, natural cement was first made in 1818 by Canvas White; the first portland cement was made by David Saylor in 1872. One of the first uses of natural cement, still manufactured in the Hudson River Valley, was for the Erie Canal. In 1916, because there were so many grades and differences in cements, the Portland Cement Association was formed to help make the products of each company consistent with others in the industry. Since that time the Portland Cement Association and the American Concrete Institute have worked long and hard to refine the product and its uses.

Everything has to have guidelines for usage. Original designers found that the strength of concrete increased rapidly for the first 15 days after it was poured; then the strength-time curve flattened out gradually until, at 28 days, the line was almost horizontal. That is, the gain in strength was appreciably great for the first 28 days, but after the initial 28 days the compressive strength increased little more. For this reason, concrete designs are made and concrete is specified on the basis of compressive strength at 28 days.

Engineers also discovered that the strength of normal portland cement concrete reaches, approximately, 60 to 70 percent of its 28-day value in 7 days. If the student wishes to work with a more exact figure he could use the following formula:

$$S_{28} = S_7 + 30\sqrt{S_7}$$

where S_7 is strength at 7 days and S_{28} is the probable strength at 28 days. But, for all intents and purposes, the 60 to 70 percent figure is sufficiently close.

The strength of concrete, as we shall consider later, depends upon many things, such as the amount of cement per cubic yard of concrete, the amount of water per sack of cement, and the strength and careful blending of the aggregates used. Also, for more than 20 years, *admixtures* (i.e., chemicals added to the mixture) have become increasingly useful. These admixtures can be very important and will be the subject of discussion later in this chapter. In addition, there are cements (called **high-Early** cements) that, when correctly blended with the design mix, help the concrete achieve a greater proportion of its 28-day strength in 7 days. High-Early cement costs more per bag. However, in cold weather when newly poured concrete must be protected with heat until it cures sufficiently, the extra cost of the cement is overshadowed by the savings in temporary heat and in the fact that forms may be stripped sooner, thus reducing the amount of reusable forms required when forms are left in place longer.

16–3   Design of Concrete Mixes

Consider a barrel filled with balls whose average diameter is 3½ in. Even though the barrel is filled to the top, there is room for a considerable amount of smaller marbles, beads, and water. This material could be placed in the spaces (or voids) between the balls; even then, the barrel would not overflow. Thus, if we took a barrel of 3½-in. balls, a half-barrel of marbles, and a quarter-barrel of beads, and thoroughly mixed the whole batch, the final product would probably fit into the same barrel. The restrictive word "probably" is used because the relative size of the components is important when you are filling barrels with balls and marbles, and is even more important when you are choosing aggregates for concrete mixes.

Basically, however, the theory of the barrel of balls is the basis for a good concrete mix. The strength of a concrete mix depends upon careful varied **gradation** of large and medium aggregates (gravels or lightweight substitutes) and the fine aggregate (sand). The strength of the concrete depends on correct gradation as much as it depends on the amount of cement in the mix. Or, more correctly, all the cement in the world will not make a strong concrete mix unless the *aggregates* are well chosen and well blended. Then the cement content is important.

In a good concrete mix the large voids should be completely filled with smaller, gradated aggregates. Then the voids between the smaller aggregates should be filled with sand. Finally, the voids between the grains of sand and between the sand and the (large and small) aggregates must be filled with a paste of portland cement and water. The reader will understand that, if the voids between the 3½-in. balls gave a maximum distance between the balls of ½ in., then a ⅝-in. or larger marble would cause the 3½-in. balls to spread or bulge away from each other. This is equally true in concrete mixes. The voids between the large, strong aggregate should be filled by smaller aggregates and sand, but **not overfilled.** A piece of medium aggregate that is larger than the void it was intended to fill tends to make the mix "roll." Finally, after the larger and smaller aggregates have been blended together, each grain of sand and each piece of aggregate must be coated with the portland cement–water paste, which must, in addition to coating each piece, fill any minor voids not filled by sand.

16–4   Natural and Man-Made Aggregates

There are two basic types of large and medium aggregates. There is gravel excavated from hills or banks and called (as excavated) "bank-run" gravel. It is then separated into large, medium, and small gravel, and sand. The gravel procured from these sources is not always round, but it tends to have rounded edges, which help the larger aggregates blend and mix with the smaller gravel aggregates.

Another source of aggregate is large stone, which is placed into rock-crushers and broken into different sizes as required. Obviously, if aggregate is to come from the breaking of larger rocks, there will be few, if any, rounded edges such as one finds on gravel aggregate. Thus older design specifications gave preference to gravel, because a concrete made with rounded aggregate tended to flow more easily into column and beam forms and the forms for basement walls. Gravel-aggregate concrete was preferred for any use where there were tight conditions in forms or considerable reinforcing steel. By the same token, crushed-stone aggregates were often preferred for concretes to be used in roads and large mass footings where pouring was easy and not a problem. Now, however, with the advent of admixtures, concrete laboratories are designing excellent crushed-stone mixes that flow well and can be used in many situations where clearances are tight.

In addition to the *natural* aggregates found in pits or crushed from larger rocks, there are aggregates that are by-products of industrial processes (such as the slag from the manufacture of iron and steel), and lava-like aggregates from natural sources, which are mined from the earth, crushed to sizes, and then processed for the concrete industry. Slag, lava-like aggregates, and "vermiculite"-type aggregates are much lighter than stone aggregates. Also, many are porous and tend to absorb some of the cement into their exterior surfaces. This tends to give the concrete more shear strength, but they may have slightly less compressional strength than natural aggregates. Thus slight mix redesign is sometimes necessary.

Most lightweight concretes use the slag-type man-made aggregates because of the lightness and strength of these aggregates. They produce a fairly strong concrete with a weight of approximately 110 lb per ft^3, as opposed to stone-aggregate concrete, which weighs approximately 150 lb per ft^3.

The vermiculites and similar aggregates are used for concrete where less strength is required. Such uses would be for roof decks, roof fill, and fill between slabs.

16–5   Sand for Concrete

All specifications for concrete-mix sand require that the grains of sand be rough and have sharp facets. One can determine that concrete sand has sharp edges by checking sand from a particular source microscopically in a laboratory, or one can blindly use it. If the grains of sand are smooth and round, there will be "placement problems" for the cement finishers.

Concrete sand comes from bank-run gravel pits or from the bottom of the sea. In the New York City area a most popular sand came from the bottom of "Cow Bay." Long after most of the good concrete sand had been dredged from the bottom of Cow Bay, concrete plants were using sea sands *like* Cow Bay sand, and designers continued to specify Cow Bay sand, fully knowing that the sand used in their concrete would be from a different location, but would produce a comparably good, fine aggregate.

Although sea sand is used in concrete, beach sand is not. Having been constantly rolled by the surf, beach sand is finely polished and rounded. One might think that this would make the concrete mix flow more easily, but it doesn't. Even if a beach-sand concrete is used for something easy to place like a sidewalk, it "just lays there," and cement finishers have considerably more work in placing and finishing it.

Quite often sand that comes from bank-run sources does not contain enough fine grains, and if the **fineness modulus**[1] of the sand is not correct,

[1] Fineness modulus is the ratio of the fine sand grains to the larger grains of sand in the fine aggregate mixture.

more cement is required. At such times, the laboratory that is formulating the design mix may require that the plant buy **mason sand** (a finer sand) and blend it with the plant's own sand. Again, *good gradation* is important.

16–6   Water–Cement Ratio and "Slump"

Basically, concrete is strongest when just enough water is used to correctly mix the cement paste into the concrete mixture and correctly place the concrete. An old gentleman, when asked for his receipt for a Christmas Punch advised, "first you pour the whisky into the punch bowl, and then remember—every drop of water you add spoils the punch!" This does not quite follow with concrete mixes. One needs enough water to correctly mix the ingredients. But, after *precisely enough* water has been placed into the mix, every additional drop of water added weakens the strength of the concrete, and quickly, considering the ratio of strength to additional water added.

The graph in Figure 16–1 illustrates how critical the water–cement ratio

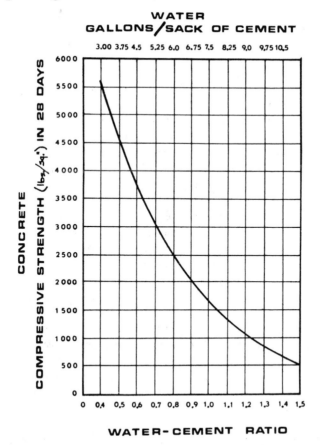

Figure 16–1.

of a concrete mix can be. Note that, **if you could** pour the mixture at 3 gallons of water per sack of cement (a sack of cement equals 1 ft³ and weighs 94 lb), your concrete might achieve a strength of 5500 psi. Probably, however, you would be pouring concrete with 4 to 5 gal per sack and would have concrete with, perhaps, 3500 psi. Consider also that, when concrete seems to be too stiff to pour and the concrete foreman on the pouring-deck whistles down to the crew at the transit-mix truck and holds up five fingers to indicate "we can't pour this stuff—add five gallons of water to the truck," at a time when the truck has only 6 yd left in it, he is ordering an extra 0.833 gal per yd, which could cut the strength of the concrete by as much as 700 psi. If the stiffness of the mix is a problem, a better way to handle it would be to contact the laboratory that designed the mix, and (with the permission of the Engineer of Record) place an admixture into each batch that will make the concrete flow better at a lower slump.

16–7   Slump and Slump Tests

A **slump** test is a quick field method of testing the workability of the concrete and, in turn, is used as a measure in checking if the water–cement ratio of the mix is (approximately) correct. Because the slump test has been used for many years, it has become a standard for engineers' specifications and a guide for field performance.

A slump test is a field measure of the amount that concrete packed into a special truncated cone settles as the cone is removed. The shape of the cone is shown in Figure 16–2. This cone is 12 in. high. It is filled to the one-third

Figure 16–2.   "Slump" cone.

level and the concrete is rodded with 25 strokes from a tapered rod to settle and *vibrate* the concrete. This is repeated at the two-thirds level and when filled. During this process, someone stands on the two metal arms shown at the bottom of the cone so that it does not lift during the filling and rodding process. Immediately after filling, rodding, and final "screeding" to level the concrete at the top of the truncated cone are completed, the cone is carefully lifted. The concrete will widen in the middle as the cone is lifted and the top of the molded concrete will *slump* or lower. If the 12-in. cone of cement were to lower to 9 in. after the cone was removed, the testing engineer would mark his test as a concrete with a 3-in. slump (i.e., 12″ − 9″ = 3″). Assuredly, this test does not measure the water–cement ratio of the concrete; it is a **guide,** and greater slump indicates a larger water–cement ratio. For this reason, most specifications note the maximum amount of slump the design engineer will allow, and laboratories that sieve the sand and larger aggregates to come up with the correct pounds per square inch mix for a concrete plant's materials will bear this in mind. As an illustration of the slump-cone system, Figure 16–3 shows three separate examples. Figure 16–3a shows a

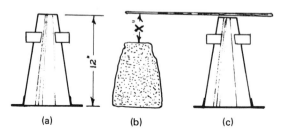

Figure 16–3. "Slump" test procedure.

12-in. slump cone as it would stand after filling and rodding. Figure 16–3b shows how the uncased concrete might look after the slump cone is removed. Because it would be difficult to measure the height of this concrete cone, the usual practice is to place the molding cone alongside the "slumped" concrete right after it is lifted. Then the rod is placed across the top of the cone and the distance below the rod to the top of the slumped concrete is measured. This is shown as X inches in Figure 16–3c.

16–8   Compression Tests

For the purpose of testing the strength of concrete, the forms or molds have a diameter one-half the height of the cylinder. Thus, originally, cylinders 8 in. in diameter and 16 in. high were put into compression machines and tested. For some time now, however, the standard test cylinder has been 6 in. in diameter and 12 in. high. This cylinder is filled one-third of its height and given 25 strokes with the tapered rod for each one-third filling, in the same manner as the slump cone was filled and rodded. The top of this cylinder is screeded off, and three more cylinders are made from the same batch of concrete that is to be tested. The cylinders carry identification marks, such as 125-A, 125-B, 125-C, and 125-D. After one or two days of storage in a safe place in the field where the cylinders would be protected from low temperatures and loss of moisture (such as a box filled with wet sawdust with a lighted 60-watt bulb in cold weather), the cylinders are taken to a testing laboratory and stored in a "moist room" with a 70° F average temperature until they are tested. During storage, the unmolded cylinders will have been capped so that their ends are perfectly smooth and exactly 90° to their axis. After 7 days from pouring (i.e., molding), cylinder

A will be placed in a compression-testing machine and an increasing load placed upon it until the cylinder fails.

If a cylinder of 6-in. diameter is used, its area will be 28.2743-in.[2]. Thus, if cylinder A (7 days old) broke at a total compression of 53,438 lb, the breaking strength of this concrete would be 53,438 ÷ 28.2743, or 1890 psi. Using the 60 percent factor, the 28-day strength of the concrete could be forecast as a safe 3150 psi. After 28 days, cylinders B and C would be broken in a similar manner. If cylinder B broke at 86,992 lb (3077 psi) and cylinder C broke at 88,499 lb (3130 psi), these two tests would prove that the 3000-psi requirement had been fulfilled. Actually, the laboratory could show the two unit strengths and then would show the average of 3104 psi for the two.

The laboratory could[2] discard the fourth cylinder, which was made as a "spare" in case the 7- or 28-day tests did not come up to anticipated strengths. In the case that these tests *did not* come up to required strength, the laboratory would confer with the Engineer of Record, and remaining cylinders would be broken later to see if added time would bring the concrete to the required strength.

16–9   Actual Curing of a Project's Concrete

Concrete test cylinders are maintained in perfect (test-tube) conditions in a laboratory's moist room. Concrete on a project cannot be maintained in such perfect conditions. However, we can endeavor to keep the moisture within the concrete and the temperature at correct levels during the initial curing period. Wall concrete does not present a moisture problem until the forms are stripped. Therefore, we should leave forms on concrete walls at least four days in normal weather and longer in cold weather. Concrete slabs should be protected from moisture loss during warm months by watering with lawn sprinklers or by spraying a curing compound (which tends to seal the surface) onto the top of the slab. The author prefers the water treatment. In colder weather the temperature of the slab concrete must be protected from cold by maintaining salamanders or other heaters under the slab. The heat *and* moisture content of the concrete slab can be protected by covering the new concrete with polyethylene, straw, and then a layer of tarpaulins to keep the straw in place. In addition, thermometers should be placed directly on top of new concrete in cold weather (under the tarpaulins) so that the temperature of the concrete may be checked from day to day. After sufficient time, and if blows from a hammer bring clear "rings," protection may be removed and forms may be stripped and the slab reshored.

[2] Current practice is to break all three 28-day cylinders (B, C, and D) if B and C come up to 28-day requirements, and average all three. If either B or C does *not* come up to 28-day requirements, D is usually kept to 72 days before breaking.

Most cities now require that a laboratory inspector monitor the loading of every trans-mix concrete truck and that a laboratory inspector take a certain number of test cylinders for each pour. If we are to be conscientious about a project's concrete, we should not add water to the truck after the test cylinders are made, and we should record the area of the pour that received the concrete at the time the test-cylinder concrete was taken, so that, in case of low "breaks," we can pinpoint the area where test cores should be taken. In addition, we should protect the concrete on the project from cold and moisture loss so that the project's concrete will achieve strengths comparable to laboratory tests breaks.

16–10   Alternative Methods for Testing a Project's Concrete

The question that always come to an instructor is "what does a design engineer do (or require) when a series of concrete tests do not come up to requirements?" First, he must try to determine whether the tests are faulty (perhaps certain test cylinders were dropped on the way to the lab) or if, actually, a portion of the structure's concrete is faulty. If the latter is true and concrete must be removed and replaced, one must remember that in 7 days (i.e., the time it will take to get the first reading) another story of concrete will be poured. In a steel-frame concrete-slab building this would be an unhappy situation. In a completely poured concrete building this would spell disaster!

Primarily, cores would be cut out of the concrete in suspect areas and these cores would be load tested. Also, floors could be load tested for deflection. Under usual (minimal) conditions, a 42-day core test may come up to a break, which although lower than the design 28-day requirement is allowable under the safety factors of the code.

16–11   Recapitulation

The strength of structural concrete depends on the mix and on the people who place and protect it until it is cured. There should be no problem in procuring correct concrete from the plant. And if someone at the plant makes a mistake on one batch, a trained concrete man at the site will usually catch it and reject that load.

However, plant mistakes are rare. Sometimes changes in aggregate sizes will not be noticed by the plant inspector and a load may be a bit "boney" (i.e., may feel to be poorly gradated). However, the cylinders will come close to requirements. Basically, whether a project's concrete comes up to design requirements is determined by the contractor who places it, protects it, and finally strips it. Here is where the Project Superintendent and other supervisors must be responsible. Here is where the success of a project is made.

Chapter 17

Studies of Stresses in Girders, Beams, and Concrete Slabs

Note that the title of this chapter mentions "studies" as opposed to complete discussion. It would be helpful if all construction superintendents and project managers had degrees in structural engineering. However, it is mandatory that the person who supervises the erection of a building at least *understands* the engineering principles of the structure he is erecting, so that he does not endanger important structural aspects and the intent of the designer and that he makes sure that the construction follows the intent.

In this chapter, or in the entire text, we cannot give instructions to the reader which will make him a structural engineer. Many texts would be required to cover merely the stress problems. However, in this text, we shall explain the important structural facts that the construction superintendent must know if he is to oversee construction methods that will produce the "design intent" of the Engineer of Record, and avoid field-construction problems that would endanger the structure. Misguided alterations of the structure or an incorrect erection of a portion of a new building may endanger the structure. These innocent "minor" adjustments or changes made by untutored mechanics are made **far too often.** Fortunately, there is a "design factor of safety" required, and thus structures are protected from most erection inconsistencies.

A structure is designed by using certain formulas and standards. Each formula can be derived and *is* derived in a structural design textbook. This text will list two of these formulas, several offshoot formulas, and will partially derive two main formulas so that the reader may understand the several charts or "plottings" of those formulas which appear in this chapter. However, we cannot take time for further derivations, and the reader is not required to dwell deeply upon the formulas listed. But, the reader is asked to *remember* the charts and their meaning. In addition, he should remember where to find the formulas at such time that he needs to refer to them. Remembering the two main formulas is not important for our needs; if the reader needs them in the future, they are given in many sources. The requirements necessitated by these formulas and charts are important.

If, at the end of this chapter, the reader understands where a beam can be penetrated for a pipe's routing without endangering a structural member and where construction pour stops can be placed so that the designer's intent is achieved and not endangered, the reader will have achieved the goal to which this chapter addresses itself. However, please understand that changes in structures or methods of erecting structures must have the prior approval of the Engineer of Record in special structural matters and prior approval of the Architect in changes to architectural details. Hopefully, all pipe and duct penetrations will be shown on the structural and mechanical drawings, and the designers will have shown any structural reinforcement required; but when an emergency arises, the construction superintendent should know how the problem may be handled.

17–1   Stresses in Steel Beams

First let us consider a steel "I-beam" in simple-support conditions. A **simple beam** is one that is supported at each end where there are no restraints producing **torque** or twisting action, as opposed to a **continuous beam,** which is supported by a number of supports that produce reversal of stresses. Furthermore, let us assume that this steel beam has a *uniform* loading (in this case a uniform loading of w pounds per lineal foot). This situation is diagrammed in Figure 17–1a. In this figure the loading system of w pounds per foot is the standard loading system of most English-language texts and engineering formulas. Whereas most English-speaking countries are now committed to a change to the metric system, this change will take a few years and will not alter the basic discussion given here.

In the discussion of loading values used to design a structure, there are loads caused by the weight of the structure, which are called "dead loads" and there are loads that will go into the structure later (such as furniture, stored items, and people), which are called "live loads." The purposes of this text are not served by discriminating between live and dead loadings. However, we *do* discriminate between **uniform loadings** (such as are indicated in Figure 17–2 and 17–5) and **concentrated loadings** (such as is indicated in Figure 17–3). The weight of the structure itself, including the beams, floor slabs, partitions, and the like, are figured into a uniform loading for the design of the entire structure. These loadings are carried by columns, and the loads from column to column are, obviously, concentrated loadings. If a certain column was to be supported at some midspan point of a beam (in this case usually called a *transfer-girder*), this would be a "concentrated loading." Also, if the engineer knows that there will be some extra-heavy object (such as a very large office safe or a large machine) on a certain floor, he will consider this as a "concentrated" load at a certain area of the structure and will strengthen the structure in the involved area(s) to carry it.

The beam in Figure 17–1a, its weight, and the weight of the uniform loading are supported at each end by two reactions, which, for now, we will show as R. Consider now that this beam will bend under the loading, as is shown *exaggerated* in Figure 17–1b. If the beam were to deflect, there

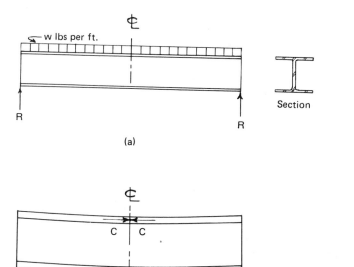

(a)

(b)

Figure 17–1.

would be a tendency for the lower flange of the beam to stretch and the upper flange of the beam to shorten. In such a case the stretching action in the lower flange would cause tension, and the shortening action in the upper flange would cause compression. The T and C symbols shown in Figure 17–1b are those usually shown in diagrams. These two figures show the situation for the simple beam.

If a **longer** beam is supported over several reaction points (a situation that is more normal), or a number of shorter beams are spliced through over several reaction points, this would constitute what is termed a "continuous-beam" system. In this case there would be a reversal of the bending action at each support (again shown as R), and so there would be tension in the **top** flange and compression in the **bottom** flange over interior columns or support points. Such a situation is shown in Figure 17–2.

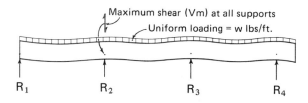

Figure 17–2.

Note also in this diagram that we have shown one end of the continuous beam extending past the support at the right, much as a beam or structural slab would extend to support a balcony on an apartment structure. This overhanging beam is termed a *cantilever* beam. The top flange will be in tension for the entire length of the cantilever from the point that it passes the support, and the bottom flange will be in compression for this same distance.

Now consider one more factor that is more easily visualized in Figure 17–2, which shows the continuous-beam situation. At every point where the beam (or slab) crosses a support there is a tendency for the beam (or slab) to shear itself off. For simplicity we have diagrammed this stress at reaction 2. However, one should understand that there will be maximum shear at all support points, and that there will be varying amounts of shear along most sections of beams and slabs, just as there will be varying amounts of tension and compression along the length of beams and slabs. The **shear**[1] will be in proportion to the shear diagrams, and the tension will be in proportion to the moment diagrams for the appropriate conditions.

Up to this point we have merely discussed tension, compression, and shear, along with the loadings and moments that produce tension and compression and the loadings and reactions that produce shear. Now we shall discuss *how* these stresses will increase and decrease along the length of the beam or slab. The understanding of the location of maximum and minimum stresses is the basis for the next portion of the chapter and is one of the most important bits of knowledge that a construction man must have. Study well!

17–2   Location and Variation of Stresses

We have previously described shear, which will be shown as *V* in future diagrams. We must also describe **bending moment** or **moment**. Moment is a product of a force times a distance. Thus, if a mechanic used a 2-ft-long wrench on a bolt and exerted a force of 10 lb on the end of the wrench, he would put a force of 20 foot-pounds (10 × 2 ft-lb) moment or torque onto the bolt. The dictionary describes *total moment* as the algebraic sum of all the rotational forces about a given point, but because we are not going very deeply into mathematics or serious structural engineering studies, we need a more simple explanation. Although the explanation given is correct and may be helpful to the reader later, let us consider the rotating forces on the beam shown in Figure 17–3. And, for simplification, let us consider the total weight of the beam and any uniformly distributed load as the **total**

[1] Shear is a force that is at right angles or at least from a side direction, which tries to shear off the member. For instance, the downward force of a beam end onto the bolts that hold the beam to the girder is the shear force on these bolts. Moment is defined in Section 17-2.

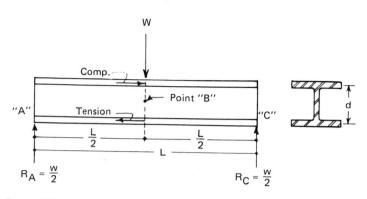

Figure 17–3.

force W, which, for this explanation (showing this loading as concentrated is not absolutely correct, but it is often so shown for discussion purposes), will be concentrated at the midpoint of the beam. Let L be the length of the beam, and d the effective depth of the beam.

If the total weight of the beam and its uniformly distributed loads were W pounds, the two supporting reactions at the ends of the beam (points A and C) would each be equal to $W/2$ pounds. The load or downward forces of W pounds would be equaled by the sum of the upward reactions (i.e., $W/2 + W/2$).

However, if one considers the *rotational* forces around midpoint B, we have one reaction at point A twisting the assembly in a clockwise rotation with a torque of $W/2 \cdot L/2 = WL/4$.[2] We have a counterclockwise rotation (caused by the reaction at point C) equal to $W/2 \cdot L/2 = WL/4$. Thus the two moments equal each other. Note also that we have shown only half of the tension and compression symbols (as opposed to the symbols in Fig. 17–1b) for simplicity. The reason for this simplification will become apparent shortly.

If the beam is steel or wood (not having extra reinforcing at one extreme as would a concrete beam), the compression in the top flange (or top half of a wooden beam) would be equal to the tension of the bottom flange (or bottom half of a wooden beam). The steel beam lends itself to simple explanation, and if we consider the torque strength of the web as a portion of the t and c stresses in the bottom and top of the flanges, respectively, the explanation is further simplified. If one takes the resistive moment of the flanges around point B, which is the midpoint of the span and the midpoint of the section's depth, then the beam's horizontal forces (s which is equal to t or c, the stresses in the bottom and top flanges respectively) will have to resist the moments (or torques) caused by the vertically placed moment of $WL/4$. Thus, as shown in Figure 17–4:

[2] L is the length of beam, and $L/2$ is a **half**-length of beam (such as indicated in Figure 17–3).

140

Stresses in
Girders,
Beams, and
Concrete
Slabs

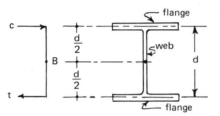

Figure 17–4.

$$c \times \frac{d}{2} + t \times \frac{d}{2} \text{ is the resistive force}$$

and because $s = t$ or c, the resistive torque or moment is s times d, or sd.

Thus, if the total W load was 4000 lb and the length of the beam was 20 ft, the moment around point B caused by the load's reactions would be $(W/2 \cdot L/2)$, or $2000 \times 10 = 20,000$ ft-lb. This moment would usually be expressed as 240,000 in.-lb (i.e., $20,000 \times 12$ in.).[3] If the effective depth (**d**) of the I-beam was 20 in., the resistive force at the beam's (exterior) flange(s) would have to be 240,000/20 or 12,000 lb in tension or compression.

From this point on we shall show our beams as truly horizontal (i.e., no deflection) and we will show the uniform (per lineal foot) loading as a hatched space on top of the beam. As opposed to the uppercase W for total load, we shall use lowercase w to indicate w pounds per *lineal* foot. The length of the beams will be a lowercase l. The shear and moment diagrams will be indicated over the beam diagrams as solid lines. The moment at any point x of the beam will be shown as M_x, and the shear at any point x of the beam will be shown as V_x. Thus, for a uniformly loaded beam the diagrams are as in Figure 17–5 for moment and Figure 17–6 for shear.

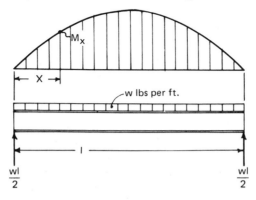

Figure 17–5. Moment diagram for simple beam.

[3] L, or l, is in feet in most cases. In final structural cases it is converted to inches.

If we have a uniformly loaded beam such as is diagrammed in Figure 17–5, the total weight will be wl, and each of the two reactions will be $wl/2$. Then the formula for the moment at any point x on this uniformly loaded beam is

$$M_x = \frac{wx}{2}\,(l - x)$$

Thus, when x is equal to 0,

$$M_x = \frac{w \cdot 0}{2}\,(l - 0) = 0$$

When x is equal to l, then

$$M_x = \frac{w \cdot l}{2}\,(l - l) = 0$$

The **maximum** moment is found in the center of the beam (or span) when $x = l/2$. When x is equal to $l/2$,

$$M_x = \frac{w \cdot l/2}{2}\left(l - \frac{l}{2}\right)$$

$$= \frac{wl}{4} \cdot \frac{l}{2} = \frac{wl^2}{8}$$

The formula for M_x and moment at midspan can readily be found in many texts and handbooks. Thus it is not important for the person in the field to memorize them. However, it **is** important to remember the curved moment diagram. This will help in construction work.

Shear is figured on a similar basis. In Figure 17–6 we show the diagram for the same uniformly loaded beam with the shear diagram above it. From

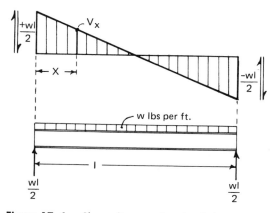

Figure 17–6. Shear diagram for simple beam.

the diagram shown in Figure 17–6 it is obvious that the maximum shear on a uniformly loaded beam is at the ends of the beam or support points (i.e., at the columns). However, to follow shear as we have moment, we start with the formula for shear (V) at any point x on a uniformly loaded beam:

142

Stresses in
Girders,
Beams, and
Concrete
Slabs

$$V_x = \frac{wl}{2} - wx \quad \text{or} \quad V_x = w\left(\frac{l}{2} - x\right)$$

Thus, when x is equal to 0,

$$V_x = w\left(\frac{l}{2} - 0\right) = +\frac{wl}{2}$$

When x is equal to l,

$$V_x = w\left(\frac{l}{2} - l\right) = w\left(-\frac{l}{2}\right) = -\frac{wl}{2}$$

At midpoint, when x is equal to $l/2$,

$$V_x = w\left(\frac{l}{2} - \frac{l}{2}\right)$$
$$= w(0) = 0$$

Here again, it is not important to remember the formula but, rather, to remember the shape of the shear diagram.

Now we should know enough about moment and shear to discuss **continuous** beams as pictured in Figure 17–7. This figure shows a beam which is a continuous member supported by four reactions. In actual field practice, this continuous member could be a member of concrete beams cast monolithically at supports (columns) so that twisting action (moment or torque) would be transmitted from one section to the next. The moment diagram for such a situation is shown in Figure 17–7.

The moment diagram of a continuous beam differs from that of a simple beam because the continuity of the beam will allow a transfer of bending moments over each support. In Figure 17–7 we have the symbol m over supports R_2, R_3, and R_4 to indicate that there is a moment at these points.

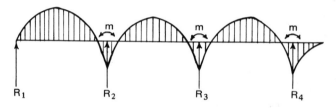

Figure 17–7. Moment diagram for a "continuous-beam" system.

We could have set a formula like $wl^2/12$ for the end moments and a
formula like $wl^2/24$ for center moments. However, in this text we do not
wish to dwell on formulas for the student to remember but, rather, **diagrams** to be remembered. Also, in actual situations where there are several
concentrated loadings along with continuous loads, formulas will vary. The
important thing to remember is the diagrams, which show *where* the
greatest moment or the greatest shear will fall.

At reaction R_1, where we have shown no construction connection to
resist moment (i.e., like the situation at the end of a simple beam), the
moment will be zero. This moment will increase as we move to the right.
However, because there is a resistive moment at reaction point R_2 (due to
continuity), the point of zero moment will not fall over R_2 but to the left of
R_2. At reaction R_2 there will be a negative moment (i.e., one that will
produce tension at the top of the beam). At points where the moment line
crosses the horizontal graph line (i.e., to the right and left of reactions R_2
and R_3 and to the left of R_4), there is zero moment. Where the moment line
is above the horizontal line (called **positive moment**), there will be tension
in the bottom fibers of the beam or slab and compression in the top fibers of
the beam or slab. Where the moment line is below the horizontal line
(called **negative moment**), there will be tension in the top fibers and compression in the bottom fibers of the beam or slab.

The beam has no end support after it passes reaction R_4. Thus the
support of this "cantilever" section will have come from R_4 and the beam
to the left of reaction R_4. The vertical support will come from the vertical
reaction at R_4. The resistance to torque or moment will come from the
continuity of the beam. And, in steel or concrete, torque or moment may be
resisted by the "**hard**" connections that resist moment, and place a part of
the moment into the column system. Thus, at the cantilever point and at
other reaction points (or columns), some moment may be taken by the
column system. In the case of the cantilever portion, there will be a negative
moment along the *top* of the beam for the entire length of the cantilever.
And there will be tension along the entire length of a cantilever beam at the
top and compression along the entire bottom of the beam. The amount of
each will be proportional to the moment curve, as shown in Figure 17–7.

The shear diagram for the same continuous-beam system is shown in
Figure 17–8. Compare this with Figure 17–6, which diagrammed the shear

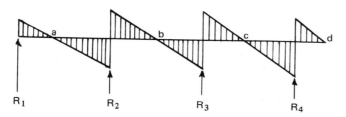

Figure 17–8. Shear diagram for a "continuous-beam" system.

for a simple beam. In the simple beam the maximum shear came at the supports, and the point of zero stress came halfway between the supports. However, when there are more than two supports for a beam system, the center supports may take a greater proportion of the loading than the end supports. Thus the points where the shear line crosses the horizontal (i.e., where the shear is zero) may not be at the exact center of the beam or slab (note, however, that in a center building situation, such as between R_2 and R_3, where reactions would, normally, be equal, the zero point b **is** at the midpoint). Zero point a will be to the left of center because reaction R_1 would (usually) not have loading as great as R_2. Reaction R_4 must carry half the load between R_3 and R_4 and, in addition, all the cantilever's load. Thus there will be more shear at reaction R_4; therefore, zero point c falls to the left of the midpoint.

Nevertheless, note that **regardless** of special systems the zero shear (points a, b, and c) falls within the *middle third* of each span when spans are approximately equal. Unless the Design Engineer has some special concentrated loadings to support, this will be the normal situation. It is important that the construction man remember that, under *normal* conditions, maximum shear is at the support points (columns) and zero shear is somewhere within the *middle third* of the span. He should also remember that maximum positive moment falls very close to the midpoint of the span, and that maximum negative moment is at the interior support points. About this time we hope that the reader is asking himself "why should these facts be remembered?"

If the designer of a building knew exactly where every pipe would be in the building, he would show every location where the pipes were to penetrate the beams and, where extra reinforcing was necessary, he would show the amount and location of the extra reinforcing. Under optimum situations, this would be the case. However, even under optimum situations, field conditions may require minor adjustments. Also, if the designer of the building knew what proportion of the concrete slab the contractor would pour, he would show pour stops on his design drawing. However, even if he locates the pour stops, there will be a day when emergency, such as a breakdown in the concrete plant, will require that the superintendent place an *emergency* pour stop that differs from the original plan. For this reason, the superintendent must be familiar wth design theory.

To this point we are discussing steel beams. Whereas the greater portion of the moment in a beam is found at the mid-portion of the beam, most of the moment is taken by the top and bottom flanges. Thus a hole through the web of the beam would be more acceptable if it were placed in the middle third of the span where the shear is minimal. If the hole has to be closer to the support points (where shear is greater), the web of the beam might have to be reinforced with extra plates (or "stirrups" in a concrete beam) to give the beam an equivalent cross section to resist this extra shear. Hopefully, all required penetrations will have been coordinated by the mechanical and structural designers, and the holes and any required reinforcement will be in the beams when they arrive on the project. However, if the location of a penetration changes, the Project Superintendent should know where the relocation *can* be placed before he asks for approval by the Engineer of Record. This engineer would prefer that any penetration of a beam be at the midpoint of the span and at the neutral axis (or the center of the beam where shear is zero and tension and compression are zero). However, he can make adjustments more easily if we stay away from the support (column) points. And, this engineer would prefer a "stop" between concrete pours to be within the middle third of concrete slabs and beams.

As we continue this discussion on moments and stresses and move on to concrete structures or concrete slabs on steel framing, we shall continue with the theories of *location* of stresses. We shall also discuss extra reinforcing bars, the need for them, and optimum points for pour stops.

17–3   Stresses in Concrete Structures

Basically, concrete is a material with excellent compressive strength but minimal tensile strength. Thus, in a concrete beam or slab, the compression can be taken by the concrete, but the tension must be taken by steel reinforcing bars or mesh placed in the forms. If the moment diagram shows that the tension is at the bottom of the beam or slab, this reinforcing will be placed in the bottom. However, as the construction comes closer to the columns where (in continuous-beam construction) the tension is at the top of the beam or slab, the reinforcing will be placed near the **top** of the concrete. Also, because shear is greater at the columns, extra reinforcing bars will be placed near the columns to resist this shear.

Because a concrete beam must have steel to resist tension, a simple concrete beam might have reinforcing placed as shown in Figure 17–9, section *AA*. The simple concrete beam pictured in Figure 17–9 is not complete. Certain details, which we shall discuss later, are omitted for the sake of simplicity. Basically, this concrete beam of rectangular section should be

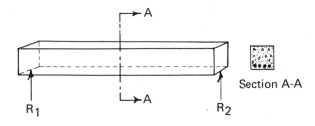

Section A-A

Figure 17–9.

compared with the steel "I-Beam" shown in Figure 17–3. Both beams are **simple** beams because their ends are not held from rotation, as would a continuous beam, which would be restrained at support points. Thus there is only tension in the lower portion of the concrete beam and only compression in the upper portion. Inasmuch as concrete does not have the ability to withstand tension, steel rods (the number to be determined by the moment in the beam) are cast into the bottom portion of the beam. The rods will be "deformed" (i.e., there will be annular rings or patterns formed onto the rods) to help them bond to the concrete. The rods in the bottom of the composite (concrete–steel) beam are sufficient to handle the tension stresses. However, if there are shear stresses beyond the strength of the cross section of the concrete and area of the steel rods, more steel rods, called "stirrups," would be added to the beam. As we progress further, we shall discuss actual concrete beam construction where these rectangular, hoop-like, reinforcing-steel stirrups are added to provide extra shear resistance. These would be added to the concrete beam where shear was greater than is allowed for concrete alone.

Figure 17–10 shows two views of a concrete beam–slab situation. Both have been considerably simplified so that the *intent* of the theory will not be clouded by showing all the actual reinforcing that would be in such a system. However, the reinforcing shown is typical. The upper diagram shows a section through a concrete beam and, to the right, a T section, which includes a portion of the slab. This is a normal configuration used in design. However, were we to try to show this T section in the lower isometric sketch, the extra drafting would complicate the isometric diagram.

In the upper diagram we have shown how some reinforcing rods could be bent upward from the bottom of the beam (in the center portion) to the top of the beam in the end portions. These reinforcing rods are bent into the upper portion of the beam because (in the continuous beam shown in the upper portion of Figure 17–10) there would be tension in the upper portion of the beam over support points. In the upper diagram we have shown cross sections of girders that would fall at column points. In both diagrams a *minimum* amount of reinforcing steel is shown for reasons of simplification.

The isometric portion of Figure 17–10 shows how the bent reinforcing is continued over the support points to serve as tension steel and to help resist shear. In addition, note that there are steel hoops or "stirrups," which are added to provide extra shear reinforcing. As there is greater shear at support points, these stirrups will be closer together near the support points, and the spacing would be increased toward the center of the beam where less shear occurs (see the shear diagram of Figure 17–8). Whereas there may be stirrups along the major portion of such a composite beam and at closer intervals where shear is greater, they are pictured (only) in the portion B of these diagrams. The remainder are omitted so that other important features will show more clearly.

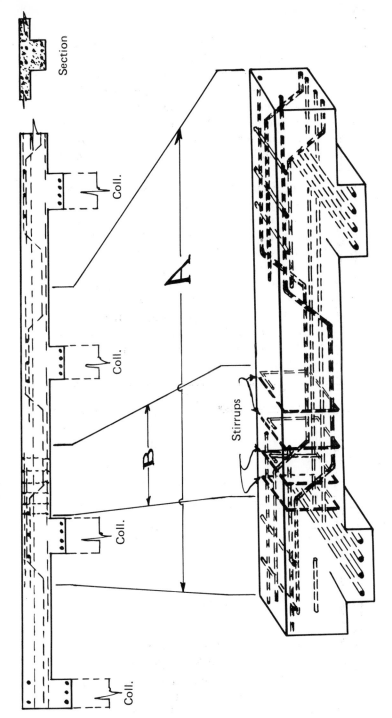

Figure 17–10.

147

When structural steel shapes are used for columns, girders, beams, and purlins of a building, concrete is usually used for floor and roof slabs. If slab loading is great, the reinforcing system of the slab will be formed with **reinforcing bars**, as shown in the upper portion of Figure 17–11. However, if the loading is light or if the spacing between the cross beams or purlins is less, then steel **mesh** may be used for reinforcing as shown in the lower portion of Figure 17–11. Either of the reinforcing materials will be placed in bottom portions of the slab in center areas and in the top portion over support points (beams or purlins).

Note that in the lower portion of Figure 17–11 we have shown concrete poured around the beams. This is often done for fireproofing of structural steel members, although most steel structures in these times are fireproofed with stray-on fireproofing systems, which we shall discuss in Chapter 19.

Whereas the concrete slab system shown in the upper portion of Figure 17–11 might be formed with plywood (which is stripped or removed after the concrete slab has cured sufficiently and is strong enough), corrugated- or cellular-type steel decking systems are often utilized these days. These decking systems are used because they replace wood forms, provide additional strength to the floor system (so that less concrete is necessary), and often provide **raceways** (conduit-like channels) for electric wiring to deck circuits. See Figure 12–1 for an illustration of this system.

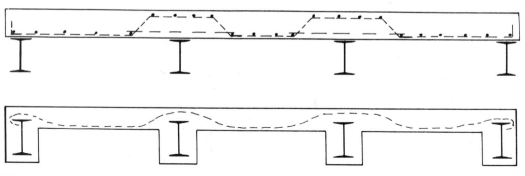

Figure 17–11.

17–5   Recapitulation

Throughout this chapter and text we remind the student that the structures he will be helping to erect will be designed by Registered Architects and Registered Engineers who have placed their seals and signatures upon the design drawings. These designers are responsible for the adequacy of the design and the contractors are responsible for erecting the structure in accordance with the approved drawings and code requirements. Therefore, if certain existing conditions require even minor changes, the contractor

should ask permission to make these changes; to protect all interests, this permission should be given in writing.

Also, because most floor slabs are too large to conveniently pour in one operation, someone must decide where to make the pour breaks. The contractor, who knows the capabilities of his crews, can best decide where it would be more convenient to place these pour stops. Having made this decision, he should mark the location of pour stops on two copies of the floor plan and submit both copies of the marked-up plan to the designer, with the request that the designer approve the system and return one copy to the contractor with his approval stamp.

However, as we noted in Section 17–2, sudden emergencies may cause the Project Superintendent to change the location of a pour stop. He should use the theory discussed in this chapter to choose this location, and he should notify the Engineer of Record (if possible) in case the engineer may wish additional reinforcing rods placed at the new pour stop.

Chapter 18

The Facade or "Skin" of a Building

Most of our older commercial buildings were clad in brick masonry or in a combination of stone and brick. Sometimes terra-cotta panels were built into the face brick. Later, concrete blocks were used for backing up the face brick in lieu of the original system of **common** brick (i.e., rough brick) backup. These facades were usually 12 in. thick (or over) on a structural steel or a structural concrete framed building. The buildings that were built were often wonderful buildings. However, because this type of **skin** or facade was heavy and because costs of masonry materials and masonry workers became more expensive, new and lighter skin systems were developed. In addition to being lighter, these facades were thinner. The older masonry system produced beautiful and enduring buildings, which had one additional advantage. Because the structure was heavy, its inertia was greater. Certainly, a building like the Empire State Building deflects horizontally (in higher portions) or "sways" in a heavy wind. However, it does not vibrate back and forth in a puffy wind as a building with less inertia does.

But then, as the costs of labor and masonry materials rose, and "pencils became sharper," as is the saying in estimating departments, designers started to use the newer facades that were being developed. Forgetting the cost of the skin or thicker masonry facade for a moment, consider the savings in **space alone** when the thickness of the skin becomes 6 instead of 12 in. If a building has an exterior size of 100 by 200 ft and the exterior skin is 12 in. thick, the remaining *interior* area would be 98 by 198 ft, or 19,404 ft^2 per floor. Now then, if the thickness of the skin is cut to 6 in., the resulting interior area is 99 by 199 ft, or 19,701 ft^2 per floor. The resulting 297 ft^2 extra rentable area at, say, $15 per ft^2 per year results in $4455 extra rent per year per floor, or almost $45,000 extra rent in a 10-year period for every floor in the building. Thus, in a multistory building, designing for this extra space is well worth considering.

However, one should not forget the larger saving in structural steel or structural concrete supporting members. Many prefabricated panel facades weigh much less than thicker masonry. This allows the structural designer to use lighter supporting members. This saving in the size of facade-supporting members starts by allowing the columns and beams of upper stories to be lighter so that, even as the columns below them are lightened because they are carrying less facade weight, they can be lighter because they are carrying less *supporting* structural steel or structural concrete above them. Thus the weight saving goes right down to the footing—a lighter structure with a lighter skin.

18–1   Thinner and Lighter Facade Systems

The types of thinner and lighter-weight facade systems are many. A partial listing would include the following:

1. Metal spandrel panels backed up with masonry and insulation to window height with continuous vision glass (usually in nonopening frames) above.

2. Sandwich panels with vision glass above.

3. Skins that are entirely glass; at the vision level there is clear glass and at the spandrel level there is obscure glass. This is backed up with insulation and masonry.

4. Precast concrete facade systems with vision glass or actual operating sash.

5. Prefabricated masonry panels (see Figure 18–1), which are fabricated with brick and chemically strengthened mortar.

We should now define some terms. A **facade** is what you look at as you view a building from the outside. It is the enclosing "partition" if you will. Sometimes, when this facade is a prefabricated product, it is called the **skin** of a building. If one considers a building's framing system of columns that support beams and girders, the beams and girders that frame between the exterior columns are termed **spandrel beams;** those which frame between interior columns (or exterior to interior columns) are termed **interior** beams or girders. Thus outside beams are spandrel beams, and facade panels attached to spandrel beams are often termed **spandrels.** If these spandrel panels are glass (as is the case in all-glass facade systems), this glass is opaque or obscure glass with bands of clear (vision) glass at the vision or window level. There are advantages and disadvantages in all systems, including the new systems. But the basic advantages (even though they are often more expensive per square foot than the older, masonry systems) is that prefabrication allows the panels to be built under **controlled conditions,** and allows for inspection by the Architect or his inspector while the panels or components are still at the factory or fabrication plant. Also, the panels may be attached or bolted to the building in weather that might not allow the erection of conventional masonry. Although more costly, the skin of a building is ready as soon as the building is ready to receive it. The skin has already been inspected and approved. And, weather is not as important as it was with the more conventional systems.

Figure 18–1. (Courtesy of Masonry Systems International)

18–2   Vision Panels

In the older buildings, there were sash that opened to allow fresh, cool air to enter. When buildings became fully air conditioned, there was a trend to make sash inoperative so that tenants would not open them and spoil the air conditioning. Thus the next step was to have continuous lines of *non-opening* vision panels. Now, with energy becoming more expensive and less obtainable, the trend is moving back to sash which the tenant may open in the "borderline" months of spring and fall when the weather changes so fast daily that the building's engineer does not know whether to heat or cool the building. In these buildings we are back to a system that gives each tenant a bit more control of his own environment and, at the same time, saves the building's environmental system. This is another reason that facade designs are continually changing. Of course, architectural fads also bring changes.

Metal spandrels are usually cast or rolled sections that are bolted to the building's spandrel beams and columns. They are usually backed up with 4 in. of masonry onto which rigid insulation has been placed. The vision section frame is manufactured by the same supplier or manufacturer. These vision sections fit or "mesh" between the spandrel panels of one floor and the spandrel panels of the floor above. A cross section of such a system is shown in Figure 18–2.

You will note that there is a **condensation-collection** opening at the bottom of each spandrel panel, and **weep holes** to allow this condensation to drain out through the spandrel. This system is necessary because there will always be dead air between a spandrel made of a temperature-conducting material and an insulated space. As the exterior temperature lowers, this dead air gets colder and cannot hold the moisture it held at a higher temperature. Fortunately, as this moisture condenses and falls from the interior air, it tends to collect on the coldest surface, which, in this case, is the metal spandrel panel. Then, as the amount of condensation on the spandrel increases, it runs down to the collecting angle and flows outside through the weep holes. The inner back-up masonry supplies the required fire rating for the building, and the rigid insulation on the inside of the masonry supplies thermal protection for the building.

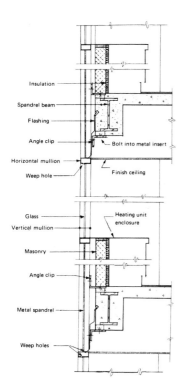

Figure 18–2. Metal spandrel panel system.

18–4　Sandwich Panel System

154

The Facade
or "Skin"
of a
Building

There are all kinds of sandwich systems in the building business. One of the most common is the board used in dry-wall partitions. This is a sandwich of two sheets of cardboard with gypsum plaster between the two sheets of cardboard, which is called **plaster board** or **gypsum board,** insofar as generic names are concerned, and is marketed under a number of copyrighted trade names.

The sandwich panel used as the skin of a building is constructed of two sheets of durable material surrounding an insulation product. The exterior material can be anodized aluminum for a high-rise office building, and could be painted or ceramic-glazed material for a motel or sales-office building where vivid color is important. Asbestos board with ceramic or other special coating is often used. The insulation product used in the center of the sandwich often determines the construction of the sandwich panel.

If a fibrous insulation such as rock wool or glass fiber is used, the panel must be constructed with angle or channel stiffeners. If the insulation is a rigid material such as polystyrene, cemented vermiculite, or one of the stronger "foams," and if this insulative material is strong enough in itself, the combination of the exterior panel material, the insulation sheet, and the back-up sheet may be bonded together under pressure to provide a panel

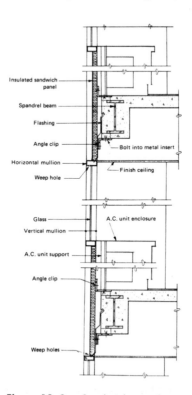

Insulated sandwich
panel

Spandrel beam

Flashing

Angle clip

Bolt into metal insert

Horizontal mullion

Weep hole

Finish ceiling

Glass

A.C. unit enclosure

Vertical mullion

A.C. unit support

Angle clip

Weep holes

Figure 18–3.　Sandwich panel system.

that will resist the wind pressure and insulate the building. Here also the sandwich panel may be of a spandrel type with a sash or vision system interlocking above it; or if the sandwich system is a metal type, the vision system will come from the manufacturing plant with the sandwich panel. In both the sandwich panel and the metal spandrel system, the vision components are field glazed, which is usually accomplished by another subcontractor.

Figure 18–3 shows a sandwich panel system. It illustrates one of hundreds of different systems, most of which are excellent. The fact that there are hundreds of systems proves that there are many different needs and the architect must choose one that fits the needs of the structure he is designing.

18–5   Precast Concrete Facade Systems

Because concrete is being precast in ideal conditions and usually steam cured to hasten and better the curing of the concrete, the process is faster. And because the system is more controlled, many beautiful architectural features such as exposed aggregate or fluted faces can be provided. These panels can be spandrel height with continuous sash or vision panels above, but usually they are larger panels that cover the entire bay (or opening between adjacent columns and the two adjacent floors) with openings for sash. This system lends itself to economy inasmuch as the panels can be cast or molded in one piece and erected in one piece. In addition, sash openings (and sometimes sash itself) may be cast into the facade section.

The insulation required for a concrete-skin system may be inside the panel itself or it may be applied later to the inside of the panels. Figure 18–4(a) shows a cross section of abutting sections where insulation has been placed within the precast panel. One must understand that all four edges of a concrete panel must be solid concrete to hold the system together and to provide strength for anchorage to the structure. Thus note that the ends of each panel have no insulation and that the end of each panel is indented or dovetailed to provide panel-to-panel consistency and strength. If this joint were to be somewhere in the middle of the bay, there would be a number of splice plates bolted onto splice bolts that were cast into the panels. If the ends fell at the column points, these splice bolts would be bolted into the column.

There is an advantage to this system in that the whole package—concrete, reinforcing, and insulation—is put together in a factory; all that is required at the site is attachment to the structure with erection bolts. There is a minor disadvantage in that the 6 to 8 in. of solid concrete that surround the panel has no insulation. If the precast concrete panel is to be covered by an internal sheathing system, this space can have bands of rigid insulation cemented over these joints. Otherwise, during cold weather periods, humid air would condense onto these narrow bands and cause considerable problems.

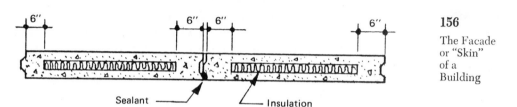

Sealant ———————/   \——— Insulation

(a)

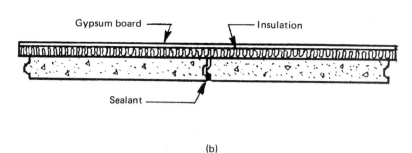

Gypsum board ———————7 |——— Insulation

Sealant ————————

(b)

Figure 18–4. (a) A cross section of abutting sections where insulation has been placed within the precast panel. (b) Here insulation has been applied *after* precast panels are erected.

If the interior side of the panels is to be exposed (as is usually the case in manufacturing plants), the precast panels could be made solid and somewhat thinner (and thus lighter), with continuous rigid insulation cemented over the entire area *after* panel erection. After the rigid insulation is installed, a gypsum-board veneer is applied, and the gypsum-board joints are taped and spackled ready for interior painting. This system is shown in Figure 18–4(b). It has the advantage that there are no "cold spots," and that the panels without insulation are somewhat lighter than the panels with insulation.

In the precast system with either integral insulation or insulation later applied to the inside face, reinforcing must be cast into the panel. We have not indicated this reinforcing because it would tend to confuse the basic diagrams. However, there must be steel reinforcing and ties in all precast panels. In the sandwich-type panels with the insulation cast into the panel, this reinforcing is usually heavy wire-mesh reinforcing, with rods and ties interspersed to coordinate the reinforcing system and to provide extra reinforcing at stress or connection points. In the solid panel system where rigid insulation is applied on the inside after erection, the reinforcing is usually bar reinforcing, because it can be preassembled in "cages" and set into the concrete forms in one piece.

Figure 18–5 shows an example of a full-bay precast concrete panel with two workable sash integrally cast right into the precast panel. The concrete for this panel is made from white portland cement, and the metal forms (or molds) had fluted shapes formed into them to create an impressive facade.

Figure 18–5. Precast panel ready for installation. (*Courtesy of Strescon Industries*)

18–6   Prefabricated Masonry Facade Panels

In our studies of concrete we noted that portland cement concrete has compressional strength, but little tensile strength. In concrete this tensile strength is provided by reinforcing steel. Thus, masonry mortar (normally a mixture of portland cement, slaked lime, and sand) has good compressional and bonding value, but should not be depended upon for great tensile strength. However, the chemical industry has invented **new** mortars. They are composed of portland cement, specially chosen sand, and chemical additives (there is no lime involved in this type of high-strength mortar). These new mortars have great tensile strength and are highly waterproof. These new mortars are so strong that within 24 hours after factory fabrication panels may be lifted by crane and set aside for final curing; when they are fully cured and have been inspected and approved by the architect's inspector, they can be taken to the project and lifted into place. These prefabricated brick panels and sections are usually only one brick's thickness (i.e., a nominal 4 in.), and they provide sufficient strength to withstand normal building-design loads, such as wind pressure and temperature movement. Temperature insulation (usually polystyrene board) is applied to the interior side of the panels[1] after erection. An illustration of a prefabricated brick section is shown in Figure 18–1.

18–7   Jointing Systems

Understandably, a facade that is made up of prefabricated sections must have jointing systems between each section to keep moisture from entering the building. The final waterproofing is usually accomplished by caulking the joints with a two-part polysulfide sealant. (Sealants will be covered in Chapter 21.) However, whenever possible, it is desirable that this sealant work be a secondary, "back-up" system. That is, whenever possible, the jointing system should include compressible tapes (usually polysulfides) or other water stops that make the facade waterproof **before** an application of a caulking-gun-applied sealant. When this is not possible, the gun-grade sealant will be the only water barrier. But with new and better sealants coming onto the market every day, the jointing between panels is more dependable.

18–8   Insulating a Building

As soon as a designer departs from the **thicker** masonry facades to the thinner facades, he must consider more insulation. And these days even the heavier masonry facades should be insulated. Actually, insulation is not difficult, but it is a portion of a building where many speculative builders may wish to cut costs. The actual process of insulating a building is easy and is a process that may well save its costs in less than 10 years because of fuel saving. The need for insulation and the amount of insulation is determined by the difference of exterior (i.e., weather) temperature and interior temperature desired. For example, if the exterior ambient temperature is 40° F, and you are content to keep the interior temperature in your house at 65° F, you do not need much insulation and you need nominal heating. However, if you wish to keep the interior of your house at 75° F, you have a much greater problem. True, the difference between 75 and 40° F (35°) is only 1.4 times the difference between 65° and 40° F, or an additional 40 percent. However, more than 1.4 times the fuel will be needed to maintain the higher temperature.

Ever since Architects have been using thinner wall sections in areas

---

[1] Except in machine rooms and the like, this insulation will then be protected with gypsum board, paneling, or other architectural treatment.

where the buildings were to be air conditioned (i.e., heated in cold weather and cooled in warm weather), they and their consulting Mechanical Engineers have been trying to get owners to allow more insulation to be designed into the buildings. In some instances the designers have been successful. However, in the bulk of the commercial buildings built in the last 20 years, owners who have erected buildings on a speculative basis have not allowed the **minor** extra expenditure for better insulation, because they might not own the building for more than 5 years before selling it to another company. These owners preferred to pass the extra heating and cooling costs (by way of rental) on to each tenant. This helped the **owner's** financial problems, and the **tenant** was not aware that he was absorbing the difference in cost. However, it did not help what was an increasing **national problem.** The United States is not self-sufficient in the production of oil or gas, and the cost of imported oil has risen dramatically. Thus the conservation rules of many municipalities have now caused building-code revisions that require more insulation.

18–9   A Crisis Demands Better Design and Better Construction

The sub-title, a crisis demands better design, has been the incentive for most building design improvements since the Stone Age. Now a crisis is *forcing* building owners to allow Architects and Mechanical Engineers to call for more insulation and more "total-energy" systems.

In 1974 our government leaders advised us that we had an energy crisis. The big problem was the **balance of trade.** Regardless of price, we had money to buy fuel from other countries and were doing it! However, because we were buying so much fuel from other countries and because its cost had risen sharply, fuel purchases were causing a large amount of our balance-of-trade problems. Too many dollars were leaving our country and not enough dollars were coming back through purchases from foreign countries. This pointed out that, just as in a household, we should be dependent on our own resources and endeavor to limit our usage to our own resources.

There was another important factor—health. Ever since we began to use more fuel to heat our buildings in the winter and cool them in the summer, we have been putting **more smoke** into our atmosphere! If you are a contractor and the cost of the building you are erecting has escalated because of fuel-saving insulation and other design, be proud of your work! Government regulations and national sources of fuel are forcing building owners to allow building designers to call for better insulation and better energy-conserving packages. If this does not become a way of life, and soon, the words of a recent poet may become unfortunate fact: "I shot an arrow into the air—it *stuck!*"

Chapter 19

Fireproofing for a Structure

Did you ever try to melt a piece of steel? It is not easy! But to *remove the strength* of a piece of steel with heat—that's easy! Did you ever hold a stick into a campfire and watch the end burn? Even though the stick is burning on one end, little heat is transmitted to your hand at the other end. The principle of these two conditions is the subject of this chapter and has been the subject of endless studies in the construction industry.

The melting points of steels vary because steel is a combination of iron and other chemical elements. The combination for these steel **alloys** is varied to give the resulting product required features, such as hardness, ductility, or strength. However, the melting point of most structural steels is over 3000° F. But the melting temperature of structural steel is not what one considers in a building fire. **Strength** is the important consideration, and one should realize that a small flash fire in a building may do horrendous things to the structure's steel in relatively few minutes!

There are few pictures in the newspapers of a structure's steel after a fire. Most news media are more interested in the actual fire than in the aftermath. There have been a few news movies taken, however, that show the deformation of steel as the fire progressed. Unprotected structural steel distorted almost as quickly as a marshmallow deforms after it is accidently dropped into a campfire.

Unprotected structural steel will probably not *melt* in most building fires. However, unprotected structural steel members lose a great percentage of their strength after a very few minutes in a fire. This is not true of timber-framed buildings. Whereas the timber will "support" (or supply fuel for) the fire which structural steel will not, the nature of wood is to provide its own insulation. Therefore, whereas a short flash fire might char a timber member of a truss or a timber column, the insulative nature of the wood itself would keep the temperature of the flames from the *inner* portion of the timber member, which would remain strong. Thus there may be enough unburned timber to support the structure. In a similar, steel-framed structure, the members would wilt like putty after a few minutes of intense heat from contact with flames. For this reason, insurance underwriters often give *lower* insurance rates on large timber structures than on similar structures with unprotected (i.e., non-fireproofed) steel framing. This illustrates the value of fireproofing structural steel.

Up until 1950, most structural steel was fireproofed with concrete. When the forms were being built for concrete floor slabs, the girders and beams were formed so that 2 in. or more of concrete would encase these horizontal members. And concrete forms were built around the columns supporting that level to give 2 to 3 in. of concrete around the steel of the column. Steel members, so encased in concrete, are shown in Figure 19-1a and b.

The thickness of concrete around steel members was set to give 2- or 3-hour fire protection, as required by local code or fire insurance underwriters. This "hour" rating originally meant that structural steel would not be damaged structurally by the heat of a fire for the number of hours noted in the required hour rating.

As you might guess from Figure 19–1(a) and (b), concrete fireproofing was expensive to form, and the portion of the concrete required for fireproofing added considerable cost and weight to the building. In addition, the weight of the steel members had to be increased merely to support this fireproofing concrete. Thus the extra cost of heavier steel to support the extra concrete and the heavier steel itself, the cost of forming and buying the extra concrete, and perhaps the cost of larger footings make concrete fireproofing expensive. However, it has one great advantage. It is durable! It resists damage from trades installing other systems into the building. It resists damage throughout the life of a building.

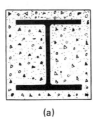

(a)

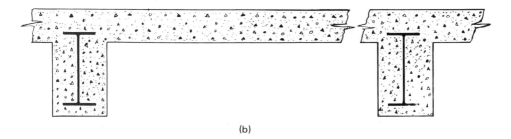

(b)

Figure 19–1. (a) Column with concrete fireproofing. (b) Beams or girders with concrete fireproofing.

Concrete fireproofing was used extensively as the *usual* fireproofing system for half of the twentieth century. However, the first departure from concrete fireproofing came in 1929 on a bank building in Dallas, Texas. Here steel members were protected by what is termed **membrane fireproofing**. These beams were encased with expanded metal lath and then coated with vermiculite plaster. The builders of this project claimed that there was a 15 percent saving in the amount of steel needed, and there was *additional* saving in eliminating a portion of the forming and stripping of the forms, plus the cost of the extra concrete that would be required for fireproofing. This 1929 application of vermiculite plaster, which is a cementitious material itself (but much lighter than concrete), paved the way for sprayed-on fireproofing (Fig. 19–2).

Figure 19–2. Spray-on fireproofing being applied. (*Courtesy of W. R. Grace and Co.*)

19–1   Spray-On Fireproofing

One of the first materials used by fire fighters in the early part of the twentieth century was asbestos. It had been used for fire-protective clothing for firemen and oil-well workers for decades, and was used in asbestos curtains in old time theaters to "compartmentize" the theater in case of a flash fire. Thus it is understandable that asbestos fibers were used in the first spray-on fireproofings for buildings. These spray-ons were a mixture of asbestos fibers, mineral wool fibers, water, and an adhesive mixed with the water and fibers to bind the fibers together and help bond the mixture to the structural steel. The first spray-ons were usually applied in generous thickness and then tamped while still damp with a board to consolidate the material. Tamping made the layer thinner but made the material more resistant to damage from operations that followed.

Soon after the largely asbestos spray-on products came onto the market, other systems were introduced that were comprised of mineral wool and a combination of ceramic or glass wool, and even wood "flour." In addition to fiber products, there were products comprised of cementitious materials such as vermiculite. These cementitious products sometimes used a small amount of asbestos fiber, wood "flour," or other material as an additional binder. However, their major composition is cementitious, as opposed to the products that are mainly comprised of fibers.

In the early 1970s asbestos was condemned for any use that allowed the asbestos to float in the air and, thus, be inhaled by workers or people in the vicinity of the work. The spray-on fireproofings that utilized asbestos were forced to remove asbestos fibers from their composition. Most fiber spray-ons that used asbestos switched to mineral wool fibers. The minor amount of asbestos used by the cementitious products was used for binder. The cementitious products thereafter used other materials in lieu of asbestos binders.

The materials used in spray-on fireproofing are placed into a hopper-type mixing and pumping machine. The fibers or cementitious materials are mixed together in the hopper, and then are forced by a pumping system into hoses, which deliver the wet material to nozzles on the floor where the material is to be applied to the structural members or to the bottom of a steel–concrete deck. Figure 19–3(a) and (b) should be compared with Figure 19–1(a) and (b), which show the original system of concrete fireproofing Figure 19–4 shows a similar spray-on fireproofing, but in this case the metal deck under the concrete–metal deck slab system has also been sprayed. This is often necessary because, whereas the use of a metal deck often allows less structural concrete, the decking material itself must be fireproofed, because the concrete–steel *deck assembly* does not, in itself, always have sufficient fire rating.

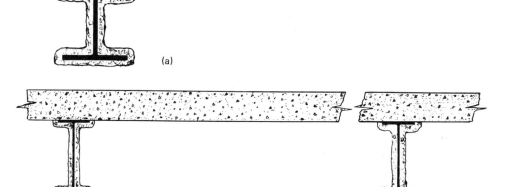

Figure 19–3. (a) Column with spray-on fireproofing. (b) Beams or girders with spray-on fireproofing.

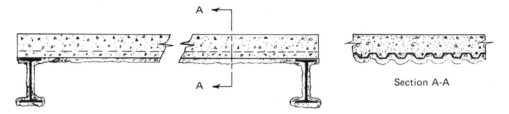

Figure 19–4. Beams and underside of metal deck with spray-on fireproofing.

19–2   Intumescent Mastic Fireproofing Coatings

Another type of fireproofing is called intumescent mastic. This material is applied to the structural members (or deck) in the form of fairly substantial paint or mastic. When flames or intense heat reach this material's surface, the mastic puffs up, forming a thicker thermal barrier. The use of intumescent materials requires proper ventilation during application and drying to minimize possible accumulation of vapors.

This system is not extensively used owing to cost and rating considerations. However, it is extensively used in small portions of construction alterations where cleanliness is important and cost is not as important.

19–3   Choosing a Fireproofing System

As noted in Section 19–2 the use of the intumescent mastic system is limited. Therefore, the main competition in fireproofing is between concrete fireproofing, fiber spray-on fireproofing, and cementitious spray-on fireproofing. Concrete fireproofing, beside being expensive to install, puts extra weight onto the structure, which in turn requires additional structural material. Thus, in these days when a lower budget is important, the use of concrete fireproofing is usually limited to those areas such as garage entrances and loading docks where great durability is important, to structural members that will be exposed to the elements for the life of the building, and to machine rooms where broken fireproofing material would cause damage if it fell into machinery. After these exceptions, the competition falls between the fiber-type spray-on fireproofings and the cementitious-type spray-on fireproofings.

Basically, the fiber-type spray-on is initially less expensive than the cementitious system. The choice depends on whether it is wise to spend a bit more money for a more durable product. Where *initial* costs are the only consideration, the fiber system should be used.

Sometimes a Construction Management organization can suggest to the Architect that a different product or type of product should be used on a project. Such an organization can do this because it is working for the owner, and it is assumed that any changes it suggests are made in the best interests of the owner. On the other hand, a General Contractor is somewhat hampered in suggesting alternative materials because he may be suspected of trying to save his company money at the expense of the owner.

Usually, the specifications of most A/Es are written to required fire rating and allow the bidding General Contractor to use *any* of the listed products or their equals as long as the fire-rating requirement is fulfilled. This system is true of most specifications for products in all divisions. However, in the spray-on division, the contractor may very well request permission to use a more expensive spray-on product if he feels that, in the long run, he will save money.

Generally, the cementitious spray-on is more resistive to damage by trades who may install their equipment after fireproofing, and it is definitely more resistive to heavy rain or water flow. If the General Contractor feels he can "backcharge" subcontractors who damage the fireproofing he may choose a less expensive fiber spray-on and then backcharge those subcontractors who damage the fireproofing (if he can prove that *they* did the damage) for "hand packing" the voids so that code requirements are fulfilled. Then, too, he may feel that he can avoid making repairs if the owner's representative does not press him.

However, if excessive water flow is contemplated during initial construction, he may very well use a cementitious spray-on and save costly repairs. A recent experience on one of the author's projects is a good example. A new pedestrian bridge between two buildings required spray-on fireproofing to the plate girders that supported the bridge and to the bottom of the steel deck and deck roof of the bridge. The bridge was to be "clad" by metal panels. Thus the structural members and the bottom of the floor and roof deck had to be fireproofed **before** the panels were installed. The General Contractor had chosen the cheapest product that the specifications would allow. Between the first application of this spray-on and the time that the sheet-metal subcontractor installed the sheathing panels, the fireproofing was washed off **three times** by rain! The fireproofing took a literal bath and the General Contractor took a financial "bath"! Had he chosen a slightly more expensive spray-on fireproofing product, one that was more resistive to damage, he would have saved considerable money.

19–4   Basis of "Hour Ratings" for Spray-Ons

When fire ratings by hours were originally established, a 2-hour rating meant that steel would be protected by fireproofing (at that time concrete fireproofing) to the extent that, after a 2-hour exposure to flames, there would be no structural damage to the steel. When the first spray-ons were introduced, they competed with concrete on the same basis. However, as more companies came into the spray-on business and as costs rose, somebody convinced fire insurance underwriters that a 2-hour rating should mean that a 2-hour fire could cause deflection in steel members, but that during the 2 hours there would be no danger to *life and limb*. This was fine for spray-on competition, but the owner of the building had to replace a lot of steel in many cases, a cost borne by fire underwriters.

Thus, about 1972, fire underwriters in a number of states made more stringent requirements for fireproofing. Many required that the application of spray-on fireproofing be under the constant supervision of a testing laboratory, which would attest to the thickness of the product. They also required that a representative of their organization be granted access to the project at any time so that he could make his own observations as to the manner in which fireproofing was being applied. To be sure that they are granted these demands, fire underwriters in a number of States have notified builders and owners that, unless these demands are complied with, they will rate a building as though the steel was **unfireproofed.**

Penny-pinching owners are often the reason for unfortunate saving measures. Thus the price of fireproofing often brings orders from the owner to the A/E to specify the cheapest materials. The requirements of the fire underwriters catch the building owner right in the pocketbook, and have been a great aid to better fireproofing!

19–5   Recapitulation

There is an old saying to the effect that "you get what you pay for." Fireproofing is usually something that is hidden from view after the acoustic ceilings, the plaster soffits, partitions, and other finishes are installed. Nevertheless, a better fireproofing material makes a structure more valuable (and perhaps safer) if the correct fireproofing system is applied. Then, too, fire insurance costs are reflective of underwriters' cost-experiences. Thus, if better fireproofing can make a number of buildings subject to lower replacement costs due to fires, eventually better fireproofing will be less costly to the owner.

From the builder's point of view, the type of fireproofing (if he has a choice in the specifications) should be chosen on the basis of total fireproofing installation costs from the time construction starts until the building is turned over to the owner. If weather or the actions of other trades are going to cause considerable nonreimbursable replacement costs, the builder should choose the more durable system. Even though costs (on everything) continue to rise, it is the author's feeling that good, durable fireproofing will eventually lower an owner's costs.

Chapter 20

Roofing and Membrane Waterproofing

The roof of a building is as important as the building itself. Throughout our lives, each of us has been advised by our elders that one should "have a good roof over one's head." Of course, this is a figurative way of saying that the base of a person's operations, the home, must be sturdy and resist the elements. No matter how well the house is built, the house is not comfortable if the roof leaks. In a commercial building, where a business must make its income, water penetration is intolerable. Thus we must always endeavor to install a roof carefully, correctly, and take pains to protect it after it is installed.

There are many types of roofing systems, and we shall describe installation methods for some. We shall cover most of the types that are used in commercial construction, give the advantages and disadvantages of some, and discuss why and where care should be taken to protect an installation.

Primarily, a roofing or waterproofing system is not a portion of the construction where designers or builders should try to save money. Yet there are some cases where this was the endeavor. The budget for a main roof or a system of roofs is a relatively small portion of the total budget of a building. For example, a recent check of a budget of a four-story, "top-quality," executive building revealed that the cost of the roofing would be only 0.68 percent of the total budget. A study of the budget for a high-rise hotel showed that the budget for the roofing was only 0.42 percent of the total budget. Thus a very small portion of the *total* building budget could be saved by scrimping on this relatively inexpensive phase of the construction. For example, if one were to try to save 10 percent of the roofing costs on a $20 million building, one might save no more than $14,000! However, once the building is finished and the owner has moved in, the cost of finding and eliminating a leak could far surpass this saving, and the possibility of damage to machinery or furnishings could far overshadow the minor savings in the original roofing system.

Regardless of the type of the roofing or waterproofing system, care in installation is most important. And because of the nature and problems of roofing workmen, it is important to understand these problems.

Primarily, we must understand that a roofer's life is not all "fun." A roofer "freezes" in cold weather, sweats more than other workmen in hot weather, breathes acrid fumes in all weather, and works under uncomfortable conditions generally. He loses time when it rains and he loses weekends when his employer wishes to recapture lost time. Thus, each year, fewer young men enter this demanding trade and there are less *experienced* men in the union hall for roofing contractors to hire. A good roofing contractor usually keeps its "key" foremen on a year-round salary and these foremen try to procure the good roofers they know when they call the hiring hall. Regardless, it is often difficult to get good men for this work; therefore, the roofing foreman, the builder's superintendent, and the Architect's representative must be constantly alert to assure a good job.

Even a good roofing team cannot do a good job unless the building contractor has provided it with clean, dry surfaces where the roofing is to be applied. If the deck surface is concrete, there must be no loose particles. When a roofing system requires **priming,**[1] it must be well applied and allowed to dry before roofing materials are applied. With the exception of the priming, it is desirable that an entire system be completed in one operation. However, when this is not possible, the roofing should be finished in sections, and each section sealed off[2] from the weather until the next working day. Also, during roofing operations and thereafter, completed surfaces should be protected from traffic. Thus, if a builder cannot finish most of the main construction work in a roofing area before he needs the protection of a roof, he might well consider installing a temporary, one-ply roof until this other construction and its foot (or wheelbarrow) construction-traffic is completed. If this is not possible, the finished roof must have (temporary) protection-sheets where this construction-traffic will occur.

20–2   Roofing Systems

All commercial roofs and noncementitious waterproofings between a structural floor and its wearing surface are one or another type of *membrane.* The dictionary defines a "membrane" as a thin, sheet-like structure

---

[1] Priming for concrete in preparation for roofing is usually an application of thinned-out asphalt or creosote, which acts much like a painter's primer.

[2] If one cannot finish off an entire roof (including all layers of insulation and roofing), he must install the roof by sections. At the end of each working day he must mop in a 12-in. strip of 15-lb felt against each exposed edge so that moisture may not enter into the system. On the next day of roofing installation, the next section of roofing will continue from this "seal-off" edge.

serving as a cover or protection. Roofing or waterproofing membranes may be one homogeneous layer, or they may be "built up" from a number of layers.

BUILT-UP MEMBRANES

Built-up membranes are constructed of 15-lb asphalt-impregnated felt cemented together (in layers) by hot asphalt or by 15-lb coal-tar-impregnated felt cemented together by hot coal-tar pitch. Note here that roofing materials must be compatible. Thus asphalt must be used with asphalt products, and coal-tar pitch must be used with coal-tar products. Most commercial roofs of this system are three-, four-, or five-ply systems with a slag or gravel topping laid into the "hot"[3] material being used. To explain, in starting a three-ply asphalt system, a roofer would start at one side of a roof and lay a 12-in.-wide strip of asphalt-impregnated felt into hot asphalt (approximately 350° F), then mop the 12-in. strip and the adjacent 12 in. of roof deck, and lay down a 24-in.-wide strip of felt. Next, after mopping the 24 in. of felt and another 12 in. of roof deck with hot, he would lay a 36-in.-wide strip of felt on top of the 12- and 24-in.-wide strips. At this point there will be three layers (or plies) at the leading edge, two layers next, and one layer for the last of the 36 in. Thereafter, using printed lines that appear on the roofing felt, he will mop in 24 in. of the 36-in. felt and 12 in. of the primed concrete deck. Laying a 36-in.-wide strip of felt into the "hot," he will continue this process, moving 12 in. (sideways) with each 36-in. layer of felt (as marked on the felt) until he reaches the opposite side of the roof, where he will lay a 24" strip and a final 12-in. strip. At this point there will be three plies of felt-and-hot over the entire roof deck. Figure 20–1 shows how these overlapping layers add up to a continuous three-ply system. It is important to note here that a full coating of hot bitumin (either asphalt or coal-tar pitch as the system requires) is necessary *everywhere*, especially on top of the primed concrete deck, so that the felt does not touch the deck directly.

We have now described a three-ply system. If a 4-ply system was specified, it could be achieved in a similar manner using 9-in. laps instead of 12-in. laps (i.e., four 9-in. layers per 36 in.), or we could use a heavier base-sheet material with 2-in. overlaps and *immediately* thereafter lay a three-ply system on top. We have accentuated the word "immediately" because, as previously noted, an entire roofing system should be done at the same time, not the base sheet one day and the three-ply another day; such a two-operation system might lead to delamination. When there is **delamination** in roofing, one or more plies or layers separate from lower plies or layers, and blister-like surfaces appear in portions of the roof's surface. Of course, these blisters will allow the roof to become damaged.

[3] "Hot" is a roofer's term for bitumin (either asphalt or coal-tar pitch) that is heated and ready for application into a roofing system. When a built-up membrane system cannot be finished off with gravel or slag until a later date, an extra flood coat of hot should be mopped onto the top membrane to seal it from moisture.

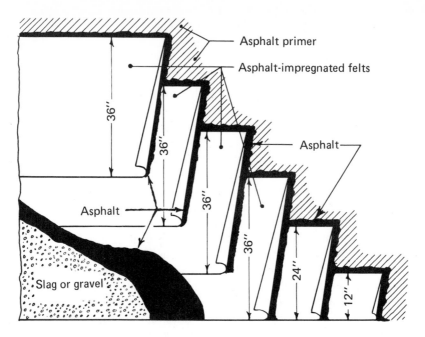

Figure 20–1. Three-ply built-up roofing system.

After all plies are completed, an additional "flood coat" of "hot" is mopped onto the membrane, and slag, gravel, or white marble chips (the latter are often used to reflect the sun's rays and lessen the heating effect of the sun) are laid into the "hot" for protection of the roof system from foot traffic and the heat of the sun. If slag, gravel, or marble chips cannot be applied until a later date, the "flood coat" is still necessary to keep moisture from top surfaces. When slag, gravel, or marble chips are applied later, an *extra* flood coat will be required.

One may ask at this point whether asphalt or coal-tar systems are better. There are scores of good reasons to choose one or the other. And there are *hundreds* of roofing-system specifiers (almost equally divided) who will defend one or the other system. It is a matter of choice. However, once the choice is made, the system must be well executed and basic materials **cannot** be interchanged.

ONE-PLY MANUFACTURED MEMBRANES

One-ply (manufactured) membranes have been on the market for the last decade. These are products of a new family of petrochemicals such as butyl and polyvinyl chloride. These membranes come in rolls or sheets to be laid directly on the surface to be waterproofed, and are either cemented to the surface with a liquid adhesive or are self-adhering. Each sheet is lapped approximately 2 in. over the last adjacent sheet.

LIQUID MEMBRANE SYSTEMS

Liquid membrane systems are those that use a liquid roofing (or elasto-meric) system, which is placed onto the surface to be waterproofed with a paint roller, spray gun, or "squeegee" depending on the liquid product and the manufacturer's recommendations. Usually 60-mil thickness in one coating is sufficient to form a durable, waterproof membrane. However, some products require a two-coat application in addition to a primer. Because of the manner in which it is applied, a liquid membrane or elastomeric membrane bonds itself to the deck somewhat as paint (which is similarly applied) bonds itself to the surface.

ADVANTAGES AND DISADVANTAGES OF THE SYSTEMS

Each of these three systems has advantages and disadvantages. All, if properly applied and sufficiently protected, will provide adequate and long-lasting water protection. The built-up asphalt or tar system is the more conventional; therefore, one finds more roofing workers familiar with correct application methods. The built-up system is usually the least expensive per square foot of the systems. Because the membrane is made up (with felts and "hot") right on the surface, the ply thickness can be varied as desired. That is, the membrane can be installed as one- to five-ply and varied on different roofs or "set-backs."

Whereas the built-up system is fine for roofs (where thickness is not a problem), it sometimes has a disadvantage when it is used to waterproof a kitchen floor that will have a quarry tile or terrazzo working surface. Here floor thickness may be a problem. In a place where thickness is a problem, the liquid elastomeric waterproofing system would be helpful and, thus, more proper.

The one-ply petrochemical waterproofing membranes, which come from the manufacturer in rolls or sheets, can fill the requirements of durability and waterproofing if they are laid onto smooth, dry surfaces and if they are afforded ample protection after installation. Because the jointings are merely 2-in. overlaps, they must be very carefully protected.

As long as a roofing or waterproofing system is *actually waterproof* there are no advantages other than cost, thickness (or thinness), and durability to damage or temperature change. However, *when there is a leak*, the built-up membranes have a disadvantage in that the water may enter at one point in the roofing membrane and exit (or become visible) at a different place. This is because, in areas where the membrane is not firmly bonded to the slab by asphalt or adhesive, the water may "travel" under the membrane until it reaches another area of the roof slab where the slab is more porous, and then become visible. Thus it is often difficult to trace the point of leakage in the membrane. However, with the **elastomeric system**, the membrane (when the roof deck is properly cleaned) bonds to the surface of the deck like paint. Therefore, when there is a hole in the elastomeric membrane, any water passing through the membrane will pass through the roof deck in the *same* area, and the leak will become apparent on the underside of the roof slab directly under the fault in the membrane. The fault is then very easy to find and can be quickly repaired. When the waterproofing membrane is to be covered with concrete, quarry tile, or terrazzo, the extra cost of the original membrane is highly overshadowed by the reduced cost of removing a smaller portion of the topping when a leak is discovered. Parenthetically, one should remember that whenever a waterproofing membrane is to be covered by tile, terrazzo, concrete, or *any* finish, the waterproofing membrane should receive a 24-hour water test with at least 2 in. of water over the highest point of the membrane before any topping is placed over the waterproofing system.

20–3   Vapor Barriers and Insulation

Whereas the heading of this sub-section is "vapor barriers and insulation," it might be more correct to head the sub-section with "insulation and vapor barriers." It's a question of "what comes first—the chicken or the egg?" *Any* roofing membrane is a vapor barrier. However, when roof insulation enters the picture, an *additional* vapor barrier is required to protect the insulation and the bond between the roof deck and the insulation.

Years ago, before we tried to keep the full heat of the summer sun or the full cold of winter winds out of buildings, we did not have a "vapor problem." Vapor is, in reality, a function of "dew point." Discussion of this subject could be taken through many pages. However, it is sufficient to say that before we air conditioned buildings, the air under the roof in the summer was relatively hot and could "hold" considerable humidity without it reaching a dangerous dew point. Likewise, the air that came from most heating systems in the winter was **dry** air.

However, with the development of insulation and air conditioning, insulations were introduced into roofing designs. When the colder air afforded by the air conditioning cannot hold the moisture in the building, or when in a plant (such as a laundry) there is excessive humidity, there is a great chance that this humidity will reach the roof insulation. Therefore, a vapor barrier is placed *under* the insulation to protect it from this moisture in the same way that the roofing membrane protects the top of the insulation (and the building) from rain. The choice of vapor-barrier system used usually depends on the roofing system. In some cases a polyvinyl chloride or polyethylene sheet is used, or a one-ply impregnated felt membrane is used. In the case of the elastomeric systems, the waterproofing membrane is applied directly to the roof deck, and no vapor barrier is needed to protect insulation from vapor below the roof deck. However, in such a case, the designer must choose a "closed-cell" insulation that will not be deteriorated by moisture from rainfall. And, in such cases, the closed-cell insulation must be protected from traffic and wind uplift by concrete or heavy gravel.

With the exception of the elastomeric or liquid membrane system, and when roof insulation is used, a vapor-barrier ply is laid, the insulation is laid next, and then a four- or five-ply membrane is laid on top of the insulation. In some cases the multi-ply membrane is supplanted by the use of a cap sheet (approximately ⅛ in.) plus a three-ply membrane. This system has the advantage of lower labor costs, but may have a **disadvantage** in that it presents a less flexible system.

Roofing insulations come from many products these days. Originally, the insulation used for roofs was sheet cork. Next came patented glass-foam products, glass fibers, exploded cementitious products, chemical and petrochemical products such as foams of styrenes or urethanes. Each of these has its economic, mechanical, and insulative advantages or disadvantages. The designer must make his choice. However, once an insulation is chosen, the material must be correctly handled and protected.

Like all roofing materials, insulation must be correctly stored and *must be kept dry* at all times. In fact, because this requirement is so important, many specifications state that insulation which has become wet must be removed from the project immediately and replaced with dry insulation in factory cartons. Many new insulations are of a "closed-cell" type and will not *absorb* moisture. However, because of their rough surfaces, they will *hold* moisture. If any moisture were trapped between the vapor barrier and the roofing, there would be a possibility of "steam bubbles" in the roofing system during the hotter months. Even if exterior heat did not cause bubbles, one must remember that wet insulation does not fully insulate.

20–4   Other Roofing Insulations

The insulations that we have discussed in Section 20–3 are marketed in sheets of approximately 2 by 3 ft, with the exception of some insulations that come as large as 3 by 6 ft. However, another manufactured-on-site system has been in existence for many years. This is an insulation roof fill system that is mixed in a cement mixer or mortar mixer. Its main ingredient may be gypsum, manufactured lightweight aggregate and portland cement, or a chemical compound that forms a lightweight concrete fill.

These poured-in-place insulation fills have the advantage that they may be poured to a variable thickness to provide pitch to roof drains and that they can be manufactured on the site when they are required. They have the *disadvantage* that, because they are manufactured *with water* and are installed immediately after mixing, they must cure on top of the roof deck until substantially **all** internal moisture has been dried out. Even after this curing is completed, these insulation fills must be protected from rain until the final roofing membrane is installed.

20–5   Protection of Membranes and Roofing
        Systems

174

Roofing and
Membrane
Waterproofing

Basically, because the membrane for waterproofing the floor under a kitchen area or machine-room area is a fragile thing, it should be protected from foot traffic or other damage as soon as it is water tested. With liquid membranes, a ⅛-in. bituminous protection board is usually used. With built-up membranes, a thin coating ($\pm$ 1 in.) of cement mortar is applied on top of the membrane to protect it, unless the cement underlay for terrazzo or tile is to be (carefully) applied immediately.

A roofing system must be similarly protected. Over the years the most usual system for protecting a built-up roof has been an application of gravel or slag into the final hot mopping. However, if the roof is to receive minimal foot traffic, a ⅛-in. cap sheet will suffice. Conversely, if the roof level is to be used as a promenade deck, a more protective and attractive system should be installed. This could be quarry tile, concrete paving, or even and exposed aggregate concrete paving using different colored aggregates to create a pattern.

20–6   Roofing Bonds or Guarantees

There are two types of guarantee systems that are available to protect an owner's finances or, if you will, to insure that his new roof will protect his building without additional expense to him: There is the manufacturer's "Roof Bond" and the "Roofer's Guarantee." Until the last decade, the "Bond" was more prevalent.

Whether finished roofs are to be protected by bonds or guarantees, most designers require that all the bituminous material to be used in the system be manufactured by the same company, or that the manufacturer of the basic waterproofing material give a written affidavit that the other products to be used by the roofing contractor will be compatible with his materials.

Under the **"Bond"** system the representative of the manufacturer of the roofing products periodically inspects the new roof, as it is being installed, to assure himself that the roofing contractor is doing a good job. After completion of the roof, this manufacturer gives the building owner a 10-, 15-, or 20-year bond on the installation. The most usual terms of such a bond insure that, if a failure occurs within the bonding period, the manufacturer will provide enough new roofing materials to make the necessary repairs. The labor for installing this new material is **not** included. The fee for such a bond is based on the square foot area of the roof (usually described in 100-ft^2 units or "squares") and the term of the bond. In the days when construction labor was less costly than the cost of the material, this system provided sufficient protection to the owner. However, now that labor is the greater portion of roofing costs, a batch of free roofing materials that must be installed at considerable (labor) expense may not give the owner enough economic protection. Thus the designer or specifier may elect to require a roofer's guarantee.

Under the "**Roofer's Guarantee**" system the roofing contractor guarantees to provide *all labor and material* for any required repairs or replacement for a specified period. The most usually specified period is 5 years. However, guarantees for longer periods are available. When one considers that a faulty roofing system will, usually, fail within the first year or two, one will understand why the 5-year guarantee is most prevalent. There is an extra cost for a Roofer's Guarantee just as there is an extra cost for a Manufacturer's Bond. Because the Bond charge is made by the manufacturer, the cost or fee is a separately listed item. Because the Roofer's Guarantee is given by the same company that is bidding the roof installation, the extra cost is usually "lumped into" the total bid and is, therefore, more illusive.

20–7   The Quality of Installation

No owner wishes to "collect" on his roofing bond or guarantee. There is usually no repayment for inconvenience or other material losses. Thus, if a General Contractor or a Construction Management organization wishes to insure a better chance for a quality roof, the most important thing to consider is the quality (or reputation for dependability) of the roofing contractor. If the reputation of a roofing contractor is good, many of the problems mentioned in this chapter are minimized. He will be a contractor that employs good workmen, uses good materials, installs under correct practice, and insures that his finished roof is adequately protected. We noted in the third paragraph of this chapter that it was unwise for a designer to pinch pennies in specifying a roofing system. It is equally unwise for a General Contractor or an owner to choose a roofer with a poorer reputation, even if there are savings involved.

20–8   Temporary Roofs

In Section 20–1 we noted that the builder might well consider a temporary roofing membrane if he could not keep construction-worker traffic off a finished roof or otherwise adequately protect the roof from traffic. Quite often, poor roofing conditions (such as too cold or inclement weather) are such that the installation of the permanent roofing system would be questionable. When a one-ply temporary roof is to be applied, there are two schools of thought. One is to apply a one-ply membrane and leave it on for a vapor barrier when the final roofing installation is made; the other is to apply the temporary roof so that it may be easily removed prior to final roofing operations.

As noted in Section 20–1 if an entire roof cannot be completed in one operation (i.e., one working day), it is best to apply the entire roofing system over *a portion* of the roof, seal the edges, and pick up the construction on the next good day. This is because there is a possibility of delamination between roof deck and insulation or insulation and top plies when the roof is applied membrane by membrane. Therefore, if questionable conditions persuade the builder to apply a *temporary* membrane, he should not risk the final roofing by allowing the temporary roofing to remain in the system. It does not make much sense to work hard to insure a bond between the temporary and permanent roofing systems only to have the bond between the roof deck and the temporary ply fail.

Thus, if one wishes to apply a temporary one-ply roof of felt and remove it prior to application of final roofing, he omits the primer (except at concrete pour joints and at roof drains), and uses only enough asphalt or pitch to "spot mop" on the deck. There should be a narrow application of asphalt or pitch along the pour joints and around the drains. Also, there should be full mopping along the joints of the impregnated felts and a full "flood mopping" over the entire surface of the one-ply.

Chapter 21

Caulking, Sealants, and Sealers

The word "caulk" appears too often in architectural details and specifications; because of a distinct difference in materials used to fill cracks and openings, the word should be used with care. Originally, "caulking," or "calking" was a system used in wooden boats and ships to waterproof the cracks between the vessel's bottom planking. Oakum, a tarred hemp in rope-like or cord-like strands, was forced into the joints between these hull planks and then covered with a thin layer of white lead before the hull was re-painted. Later, when painters prepared to paint the exterior of a building, they fashioned a compound of whiting (one of the ingredients of putty) and linseed oil to make what was called caulking compound. As this was a paint-like product that had no flexibility after it hardened, it became brittle and fell out of cracks if it was used in wide cracks.

21–1 Caulking Compound

Today's caulking compounds are, generally, the same generation products as the early painters devised. True, certain manufacturers have improved upon the formulas, but eventually the compound will harden into an inflexible material. Thus, it is recommended for *interior* crack filling that is not subject to the temperature differences experienced outside; if an exterior crack is less than ⅛ in., the compound will probably hold under a coat of paint. However, it is not recommended generally for exterior usage. Cracks are usually filled with caulking compound by forcing a **bead** of the material from a caulking gun. Thus the verb "caulk" is still used to describe the filling of cracks with any compound. The confusion (and, more importantly, mis-application) comes when an architectural detail has an arrow pointing to something in a crack that is labeled "caulking" when, actually, the designer knows that there will be expansion or movement in the crack and desires "**sealant.**" Thus, whereas a specification might read "caulk all exterior joints with a two-part polysulfide compound," there would be less confusion and less chance of incorrect application if the specifications read "apply a two-part polysulfide sealant in all exterior cracks." Also, architectural details should differentiate between caulking and sealant. There is a great difference, as we shall now discuss.

21–2   Sealants

Sealants are the second generation of crack-filling compounds. These compounds came onto the market in the mid-1940s and have made a great impact on the joint-waterproofing systems. The original sealants were two-part compounds that used a fairly soft elastic compound into which a hardening catalyst or hardening accelerator was mixed just prior to use. The usual system was to add the acceleration agent into the can of basic material and mix thoroughly with a mixing paddle on a low-speed electric drill. The mixed compound was then drawn into a caulking gun and applied to the joints.

The basic advantage of a sealant compound as opposed to a caulking compound is that, although it stiffens considerably after it is applied, it never completely hardens. Thus, as the joint between two panels expands and contracts with temperature changes, the sealant adheres to both panels and stretches and compresses as the panels move. This is something a caulking compound could never do. Because of this ability to continually withstand weather and moisture for many years and not lose its bond to either side of a joint, the sealant family has made the prefabricated spandrel system much more efficient. In fact, without an ever-elastic sealant compound, many prefabricated panels could not be made watertight and workable.

21–3   Types of Sealants

Every year more sealants come onto the market. Even while the printer's ink is drying on these pages, at least one new sealant will come onto the market. Thus trying to keep up with the new ones and with the claims for their abilities is a difficult job, even for a professional specifier.

Basically, however, there are the two-part sealants that are hardened by a catalyst or hardening accelerator, and there are one-part sealants that take up moisture from the air for their catalyst. From this point on, products differ by material. The first sealants were polysulfides; next came the polyurethanes and, more recently, the silicone sealants.

It is difficult to advise which type of sealant should be used in each situation. Because the two-part polysulfide sealant was on the market first, many specifiers and sealant contractors like to use it generally. However, few people will agree which is the best product for certain situations. It is the author's opinion that polysulfide sealants should be used in joints that are ½ in. or narrower, and that polyurethane sealants should be used for wider joints.

Generally, two-part sealants should be used on projects where quick-setting joints are desirable. An example would be the expansion joints between sidewalk sections where traffic could not be diverted for all day. In such a case, one would use a thinner, self-leveling, "pour-grade" mixture, rather than the usual gun-grade mixture, which is thicker in consistency as it comes from the original container. Both the one- and the two-part sealants come in tubs, as do the pour-grade sealants. Understandably, the sealants that must have catalysts mixed into them immediately before use (the two-part products) come only in tubs. Thus the size of the tubs should be chosen so that the entire batch may be used up before its **pot life** (the length of time that it remains usable) expires. Gun-grade sealants also come in tubes for smaller applications.

Silicone sealants (one-part products) do not reach the same degree of hardness as do the polysulfides and polyurethanes. These are used in conditions where a more flexible jointing is desirable, but where this softer joint will not be damaged. For example, hand-sized tubes of silicone sealant are used to waterproof around the joints of bath tubs and shower stalls.

21–4   Installation of Sealants

Because a primary necessity is that a sealant adhere to the edge surfaces of a joint, the applicator must be sure that all surfaces are clean, clear of materials that may pull away from a lower surface, and that the surfaces are dry. If the surfaces to which the sealant is to be applied are homogeneous to the base material, the applier should remove any dust and grime and, finally, wipe the joint with a cloth or brush dipped in solvent so that any grease or more difficult material is removed. After this has dried, the sealant may be applied. However, a good Project Superintendent would be wise to let sealant be applied to the joints between the lintels (support angles over sash) and the sash, and have the **painter follow** when the sealant is cured. True, painting specifications usually require that the painter wire brush any and all rust from ferrous metals before patching the priming and applying final paint. However, if the painter does not do a good job, it will not do much good to have an excellent bond between a sealant and the paint on the iron lintel where there is a paint-adherence problem. Certainly the sealant will stick to the paint and to the sash. However, the first time weather conditions cause the sealant joint to expand, the sealant will pull the outside paint off the rusted lintel, and there will be a crack in the jointing system. It would be better to allow the sealant applier to prepare the angle for his sealant work and let the painter follow. In this way the owner can hold both trades to the guarantee on their own work.

Fully as important is that the sealant joint be formed so that it **may function** as the designer required and adhere to both sides of the joint. That is, the sealant must expand and contract as adjacent materials move, and must fully adhere to these surfaces during these times of sealant stress. Thus the *shape* of the sealant joint is very important. Some specifiers require that

a sealant joint be as deep as it is wide, but most specifiers and manufacturers recommend that the joint's depth be from one-half to two-thirds its
width, but *never* deeper than it is wide, and *never* less than one half-width
in any joint where considerable movement is expected.

To explain this, we present Figure 21–1a through c, which shows three
control joints of identical width but three different depths. Of the three
situations, Figure 21-1b has the usually desired profile. Note that the depth
of each of the sealant joints has been determined by placing a back-up
"**rope**" into each slot or joint to limit the depth of sealant penetration.
Polystyrene rope is often used for this purpose. Polystyrene has an added
advantage in that the sealant will not adhere to it; the sealant adheres only
to the two sides of the actual joint. This back-up material should be one-
third to one-half wider than the slot between the materials so that, when
compressed into the slot, it will hold its depth. Above each part of Figure
21–1 is a sketch of the approximate configuration of the joint in tension
condition. Note in Figure 21–1a that the depth d is much smaller than the
width w. In such a case it is most probable that there would not be enough
length of material cemented to each side of the joint to work against the
tension; the sealant would pull free of one side or the other.

In Figure 21–1b, note that there is enough length of contact with the
two sides of the joint, and that the cured sealant may stretch and "cup-in" to
allow for the required stretching of the sealant joint without losing the bond
on either side. In Figure 21–1c the depth of joint is much greater than the
width of the joint. Obviously, there is enough length of joint to give plenty
of bond to the sides of the joint. However, the depth d is so great that the
sealant material cannot cup-in when pulled by tension. In such a case something has to fail. Depending on the excess depth as opposed to the material
strength of the sealant, it will be either a tear in the material or a break in
the bond on one side of the joint.

Finally, we must warn about moisture and temperature. If a joint to be
sealed is a vertical joint up the side of a building it is possible that at least
the upper portion of the joint will be dry enough for sealant after one or two
warm, windy days following a rain. However, for a horizontal joint in pav-

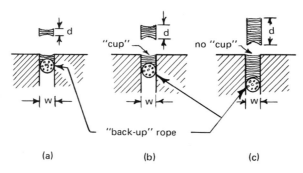

(a)   (b)   (c)

Figure 21–1.

ing, many warm and sunny days must follow a rain to be assured that the joint is completely dry. Every once in awhile some contractor feels he is accomplishing correct drying by running over a sidewalk or plaza joint with a propane torch. It does not work! And, year after year, this is proven when the sealant can be pulled out of the joint a month or so after installation.

Temperature is important also. Most manufacturers of sealants recommend that their products be installed in temperatures of 40° F or better. Confidentially, they will tell you that their product *can* cure at lower temperatures. Maybe the sealant can cure, but the concrete or masonry usually cannot be ready at lower temperatures. If masonry or concrete has been standing at temperatures of 32° F or below overnight, there will be frost inside the masonry or concrete that will seem to disappear in the morning after the sun comes up. Actually, however, although it may have disappeared on the surfaces of the joint, it is still *inside* the material. This moisture will attack the bond of the sealant.

21–5 Sealers for Masonry or Concrete

In the moistureproofing business there are dozens of liquids to spray or brush onto masonry or concrete in the endeavor to moistureproof the building material. Bear in mind that we are talking about moistureproofing, not waterproofing! If waterproofing is what is desired, forget the spray-ons and paint-ons. Such usages are only temporary and may, at a later time, preclude an application of the **correct** material.

However, for moistureproofing there are a number of liquids that may be sprayed or brushed onto a masonry or concrete surface. One of the earliest products for masonry moistureproofing was liquid silicone. This does the job but, usually, it must be reapplied after a few years. Emulsions of acrylics are usually longer lasting. Their length of service depends upon the percentage of acrylic solids in the emulsion. There are a number of moistureproofing acrylics on the market, but not all of them have enough solids. The person who applies a product with a minimal amount of solids will temporarily solve the problem, but he may be up on the scaffolding two years later redoing the job.

In the family of sealers there are, also, sealers to harden or dustproof the surface of concrete floors. Here again the amount of solids is important, and the vehicle that carries the solids must be one that will penetrate into the finished concrete floor. If the product does not penetrate far enough into the surface of the concrete, it will wear off quickly. If there are not enough solids to seal the pores in the concrete surface, there may still be dusting.

In both moistureproofing and dustproofing sealers, considerable care must be taken in choosing a material. The designer of the building should check the reputation of the product that he is specifying. If he does not, the application contractor will be wise to qualify his guarantee when bidding the work. If he complains about the product *after* bidding the work, he will be suspected of trying to avoid his contractual responsibility. Even when the sealing product is correct, the installer must take particular care when applying it.

21–6   Recapitulation

Sealers and sealants can achieve excellent results for the designer and for the builder. However, if they are to be successful, the right product must be used in the right way.

Chapter 22

Interior Finishes and Specialties: The Wet Trades

The interior construction and finishes of a building are divided between the "wet" trades and the "dry" trades. There are a number of interior finishes with wet materials that are not included in the wet trades. Even though paint is wet as it is being applied, the materials and contractors that apply paint are not listed with the wet trades. Basically, the meaning of "wet" trade is a trade that erects or installs a material which requires water in the process. Water itself is no problem or hazard in the installation of the interior construction of a building. The problem is primarily that the wet trades leave considerable wet debris. Of secondary importance is the fact that materials which have water within them cause humidity problems that may delay the installation of *other* interior finishes which cannot be installed until moisture content or humidity is low.

In the discussion of both the wet and dry trades we will cover the product involved and the manner in which it should be installed, and, in addition, the precautions that may make the problems of other interior finishes to follow less burdensome.

A partial listing of wet trades would include the following:

Interior concrete fill and floor finish (4)

Masonry (2)

Plaster (3)

Terrazzo (1)

Poured resilient floor surfaces (5)

Ceramic tile (6)

As you have read this list you have noticed a number in parentheses after each material. Is it the unit weight? Is it the order in which the product should be installed? No, it's the "nuisance factor" or the measure of the problems this material causes the Construction Management firm or the General Contractor! With this explanation you now notice that, in the author's opinion, terrazzo causes more debris problems than the other wet trades. It's at the top of the hit parade in problems but it is worth the

problems. Terrazzo is one of the best-looking of the interior finishes. However, it produces a grinding-slurry that is a big nuisance. On the other end
of the scale there is concrete fill and finish. This, too, is a product with
considerable moisture per square foot, but because the concrete fill is
brought to the area in the approximate amount needed and because the
concrete dries out fairly quickly, it leaves little moisture problems in the
days following its pouring.

However, regardless of the author's nuisance-number for a particular
wet trade, the wet trades should be completed as soon as possible in the
building construction process so that the dry trades may commence.

22–1   Interior Concrete Fill and Finish

Quite often the final finish of a concrete floor is applied along with the
concrete deck. This is called "monolithic finish." In the monolithic process,
any excess concrete is returned to the ground level the same afternoon and
therefore it does not become a debris problem. Sometimes the cement
workers place the small amount of left-over concrete onto plywood over an
adjacent pour and remove this excess the next day when they lay craft paper
over the recent pour. Thus the debris problem is quickly solved. However,
when the designers or owners of the building require a floor that has more
exact grades, or where electric duct systems must be installed later, another
layer of lean-grade and lighter weight fill is placed onto the structural slab
with a layer of finish concrete on top. Or, in many cases, merely a layer (1½
in. thick, perhaps) is poured later.

If the structural slab is to receive one or more layers of finishing concrete
later, it is roughened with a rake so that the next layer will achieve a better
bond. Obviously, an overlayer of concrete can be poured to quite exact
grades, because the structural slab is already poured and will not deflect
from the weight of the fill concrete or the workmen installing it. The most
exact manner of achieving good levels (**grades**) is to set 1-in. pipes onto
supports and, with the use of a surveyor's level, set the level of these pipes
at frequent intervals. These grade pipes are called **screeds.** Most usually the
cement finishers build small mounds of sand–cement concrete with a fairly
stiff mix, set the pipe screeds a bit high, and then tap the pipe screeds
downward slightly when the Field Engineer has his level rod set onto the
screed at the support point.

We should note here that in cases where the concrete slab is to have a
monolithic finish (i.e., the structural concrete slab is to be poured to finish
grade in one pour), pipe screeds can still be used to give fairly accurate
finish grades when they are set onto adjustable chairs, which have threaded
supports for the pipe screeds. In this case, however, the supports or **shores**
for the structural slab must be heavy enough to keep deflection of the
structural slab at a minimum.

In either case, when the concrete is poured, the cement finishers pull a
good straightedge across the top of adjacent pipe screeds to achieve an

accurate level for the top of concrete (often noted on design drawings as
T/C). As soon as the concrete hardens to the point where it will bear the
weight of the cement finishers in knee boards, the screeds are removed so
that the slot they leave can be filled with concrete and troweled in to bond
with the floor. Certain owners will have factory equipment that is set on legs
that have little adjustment; these companies require the top surface of the
concrete to be within ¼ in. of the correct grade. If the top level of the
concrete is achieved by the two-pour system (i.e., a topping layer poured
onto a previously poured structural slab) and screeds are accurately set,
such tolerances are quite possible. And such tolerances can be met on monolithic slabs, but shores under the concrete must be quite close (so there can
be no sags between them), they must be accurately set, and they must have
tight connections that will not shift after concrete is poured. Understandably, the pipe screeds must be set most accurately *after* all shoring is firmed
up.

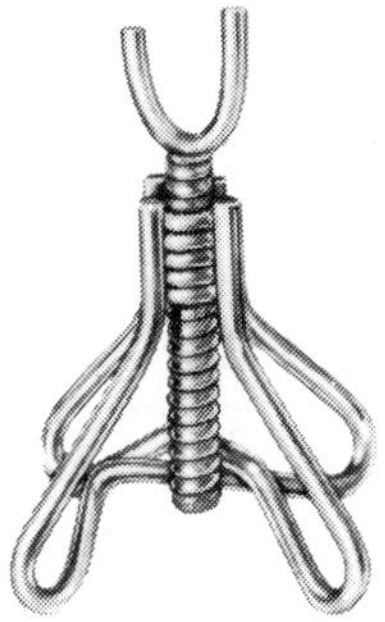

Figure 22–1. Adjustable screed chair. (*Courtesy of Richmond Screw Anchor Co. Inc.*)

Actually, if the floor-surface concrete is poured before exterior curtain
walls are erected (and this is usually the case with monolithic concrete
slabs), there is little debris problem when other interior work commences.
When concrete fill and finish is poured, the areas should be as large as
possible so that door openings (door bucks) and partitions will "find" no
level differences. For example, it is best to pour the finish surface *before* any
partitions are erected. Some companies know that partitions will never be
moved. If this is known, finish may be poured in rooms and corridors *after*
partitions (usually masonry in this case) are built. But consider, if a partition were moved at a later date and there was a ¼-in. difference in floor
levels between the two original spaces, there would be a bad tripping hazard. Thus it is best to get the floor poured before partition erection to avoid
this problem.

22–2   Masonry

Masonry is another wet trade that has a debris problem. If the facade of
the building is to be masonry, it is usually built first, and the interior
masonry partitions are built later. This procedure is followed for two rea-

sons: (1) the builder will wish to "close in" the building as soon as possible so that other trades can continue, and (2) the builder will wish to have a place to use his bricklayers on rainy days if he has a floor or two closed in. Otherwise, the men would lose the rainy days and the project would experience overall lost time.

When the masonry facade is finished at a floor and the bricklayers are working at a higher level, laborers can shovel up mortar and broken masonry pieces and pour it down a dirt chute to a debris hopper. Too many contractors delay this intermediate cleanup work until several floors are closed in. This is bad for the project. If the debris is cleaned up as a floor is passed and the floor is broom cleaned, other finish trades may start their work the next day. As soon as the floor is cleaned, the Field Engineers should lay out the partitions and use yellow zone paint (which dries in minutes) to mark important points, such as corners and partition intersections. If the partitions are to be masonry, the carpenters may then set the door bucks (frames) in readiness for interior partitions. Thus, if the bricklaying subcontractor is considerate of others, he may find that he has also served his own interests and has a place to build partitions when rains come sooner than he expected. However, if he is not a considerate subcontractor, the General Contractor or the Project Management organization must force him to clean up immediately or have the area cleaned with its own forces and backcharge the masonry subcontractor. A *"backcharge"* is a charge against a subcontractor for something he should have done or payment for an error he has caused. The General Contractor or Project Manager deducts this money from his payments. This is fully explained in Chapter 29.

After interior masonry is finished on a floor, the area should be cleaned up **again,** even if the plastering subcontractor is to follow and will, again, cause more wet debris. The masonry subcontractor should clean up his own mess so that the plastering subcontractor can work more easily in a clean area. After the plastering subcontractor finishes *his* work, he will scrape and clean the floors; thus, if the masonry subcontractor left clean spaces, there can be no chance of divided responsibility. Again, if the plastering subcontractor does not clean up to the satisfaction of the Project Superintendent, he will be backcharged for having the work done for him.

There is one precaution that all bricklayers should take and which the Project Superintendent should watch for. Mortar, like many cementitious products, shrinks as it dries. When an interior partition is built of 8-in.-high cinder-concrete blocks or regular cement blocks, a line of partitions is rarely completed in one day. Thus there is no problem. However, when lightweight gypsum blocks (which are 12 in. high) are used, the bricklayers may complete one-half the story height in a morning. The usual practice, then, is to keep the mason tenders working on overtime through the lunch hour to build scaffolds so that the masons can finish the partition in the afternoon. **This is a bad practice!** The joint shrinkage in full height of the partition may very well exceed ⅛ in. Thus the next day the partition will have no support from the slab above to keep it in place horizontally. In an elevator shaft (a place where many designers call for the better fire-rated gypsum blocks) the suction caused by the elevators may pull the partition into the shaft. Thus it is best to leave the *last course* of gypsum block to be laid the *following* day when all shrinkage has taken place. The minor shrinkage in the two mortar joints for the last course will not affect the bond appreciably. However, in the first case, the shrinkage of 12 mortar joints laid the same day would definitely cause problems.

22–3   Lath and Plaster

Because plastering (usually gypsum plastering) is a messy process in the minds of many, and because gypsum wallboard is easily erected and may be installed at a lower unit price, less and less plastered walls and ceilings are being installed as time goes on. Because of this situation, there are less and less good plastering mechanics coming into the business. But plaster does a job that no other surfacing product can do as well. Even though sculptured plaster interior cornices and "dentils" are now being produced in paper-mache, they are not the same thing. When a masonry wall is covered with plaster and when the ceiling is made of lath and plaster, the whole system is tied together and the entire structure is more rigid and more fireproof. Certainly gypsum board with insulation and paper-mache cornices will provide the same *appearance*, and usually for less money. However, there is no substitute for a good plaster job. Your author agrees with the advertising that many plastering associations have put on bus and subway cards, which read, "Keep America Plastered!"

If plaster is to go onto masonry, the mechanic works to a wood base ground and builds up a plaster "screed" ½ to ⅝ in. thick, after which he applies and trowels a coating of "brown-coat" plaster (a mixture of gypsum and sand), bringing the surface of the brown coat to within approximately ⅛ in. of the finish line. The next day the plasterer applies a ⅛-in. coat of a mixture of hydrated lime and plaster of paris to bring the coating to the finish line desired. As there is considerable plaster of paris in this second mixture, this coating hardens quickly and provides a very durable finish. If a plaster with more fireproofing ability is desired, vermiculite is substituted for the sand in the brown coat. Two 1-in. layers of vermiculite plaster will provide a 2-hour fire rating.

If plaster is to go onto wire lath, a first coat (called a **scratch coat**) must go onto the wire lath to fill the holes in the lath and to stiffen the wire lath foundation. This is usually a richer mix and is approximately ⅛ to ¼ in. on the room side of the lath and probably ¼ in. inside the metal lath. Soon after this is applied, the plasterer takes a brush-like tool made up of a dozen or more wires and scratches grooves into the "scratch coat." These grooves makes it easier for the next day's brown coat to bond to the lath and scratch coat. Curved walls are often made with lath and plaster, and plaster ceilings are always made with lath as a foundation for the plaster finish.

The accuracy of a plaster wall and ceiling depends primarily on the accuracy of the line and grade given by the Field Engineer and, second, on the accuracy of the screeds set by the plasterer.

After plastering has been completed, fans should be placed in the areas to provide moving air to accelerate the drying process of the plaster, thus allowing the painter to start sooner. There are two precautions to be observed here. Although heat will help in the drying process, the *moving air* is more important. And plaster should not be overheated. Second, and this is really a precaution for the painter, paint should not be applied to plaster until a moisture meter proves that the plaster is sufficiently dry. Sometimes it is necessary to "paint out" a space before plaster is thoroughly dry. This painting can be done (with an educated risk) by first applying a special sealer coat. However, the Project Superintendent must realize that if there is moisture in the plaster (even a small amount) he must leave a path for it to travel. Thus, when there is a chance of moisture, he should **never** order the painter to seal and paint the plaster on *both* sides of the partitions. If he does make this mistake, the escaping moisture will make blisters on one or both sides. This is because the moisture *must and will* find a way to escape.

22–4 Terrazzo

Terrazzo, a material invented in the Mediterranean countries, has a surface material that is a mixture of marble chips and a binder (called a "matrix"), which was originally a Puzzolan cement (see Chapter 16) mixture and, more recently, a portland cement mixture. Now the binder or matrix may be an epoxy or an acrylic material.

However, basically, the surface material of terrazzo is poured or cast upon a base. The marble chip–matrix mixture is allowed to cure; when the surface is ready, the rounded marble chips are ground and honed flat with rotary abrasive-capped grinding machines. This process grinds off the rounded tops of the marble chips to the point where all the pieces of marble are flat and on the plane of the honed-down binder or matrix. The result is a stone-like floor made up of small pieces of marble with a cement matrix or a matrix composed of an epoxy or acrylic material. Because there will be expansion and contraction in such a material, terrazzo is usually divided with brass, bronze, or white-metal divider strips. The top surfaces of these divider strips are also ground down in the surface-grinding and honing process.

There are now several different types of terrazzo systems. One is a thick, "floating" foundation course for the surface terrazzo; in another the foundation course is thinner and is bonded to the concrete slab; there is also the "thin-set" system in which the surface material (i.e., the terrazzo) is *bonded* to the concrete slab. The order in which we have listed the systems indicates their cost, the first being the most costly, but more "crack-proof" system.

If we knew that there would be no structural movement in a concrete structural slab that might produce cracks in the surface, we could safely cement a terrazzo surface onto this slab, grind it down, and have a beautiful floor surface that would never crack. However, structural concrete slabs (or "arches") seldom come through without minor cracks. These cracks do not pose a structural problem for the building because of the steel reinforcing. And, 9- by 9- or 12- by 12-in. floor tiles move over such cracks so that they are never noticed. However, if one bonds a terrazzo surface to a slab that has a crack or later develops a crack, this crack will show through into the terrazzo.

FLOATING FOUNDATION SYSTEM

The most expensive but safest terrazzo system is the floating system. In this system a minimum of ⅛ in. of sand is spread over the concrete structural slab. Next a polyethylene sheet (originally 15-lb. roofing paper was used) is spread onto the sand as a "bond-breaker" sheet. Then a very lean sand–cement mixture (usually a one to three mix) with very little water is laid on this bond breaker to about 3-in. thickness. In the middle of this base course there is a 2- by 2-in. by No. 14 wire mesh to reinforce this lean cement–sand base course. As soon as this 3-in. base course is leveled and rolled, chalk lines are snapped onto its surface and divider strips are set into the base to form the squares. The next day marble chips of the desired size and color are mixed with the matrix material, and the mixture is poured into the squares and then rolled to make the top surface fairly level.

If the matrix is a portland cement base matrix, at least 7 days will pass before the terrazzo may be ground and honed. If the matrix is an epoxy or acrylic base, the grinding wait depends on the base. If there is no sand base between the structural slab and the matrix (i.e., a thin-set system), the grinding and honing may begin the next day.

The 3-in. floating base with a ½- to ¾-in. terrazzo surface is resting on a roofing-paper or polyethylene breaker sheet with from ⅛ to ¼ in. of sand under it. Also, this 3-in. base course has wire mesh cast into it to keep it from cracking. Thus, if the structural arch or slab develops a crack, the crack will not come through into the terrazzo because in essence we have a floating floor system on top of the structural slab. There is no better system; the author has put in such systems as large as 20,000 ft² and, 20 years after installation, has found no crack. Assuredly, it is more expensive.

OTHER TERRAZZO SYSTEMS

The more prevalent system for years was a system which allowed grades for 1½ in. of terrazzo base course (again sand and cement with divider strips) and ½ in. of terrazzo. For this system the *structural slab* was raked to make grooves in its top surface so that the terrazzo base would bond to it. The slab under the *"floating"* system was "darbied" smooth to allow for slippage.

But the terrazzo base for this system **bonded** to a raked surface. The advantage is a durable, less expensive terrazzo surface. However, if a crack shows up in the supporting structural slab, it will certainly appear above in the terrazzo surface which was bonded to it.

Finally, there is the "thin-set" method. Here metal divider strips are fastened (by epoxy usually) to a smooth-troweled concrete floor slab. The next day a marble chip mixture with either an epoxy or acrylic matrix material is poured into the squares. After the terrazzo mixture has been rolled to level it off, it is left to the next day, when it can have its initial grinding.

There is one other thin-set terrazzo system. This is the **"conductive terrazzo,"** which allows electric currents to be dissipated from operating tables in hospitals or other laboratory equipment through to a grounding system of the floor. The system uses a neoprene-type mixture with graphite in its matrix. The system has an advantage to the person standing upon it in that it is slightly more resilient (i.e., softer) to the feet. Its disadvantage is that, because considerable graphite is used to give the conductivity, the matrix comes in but one color—basic black!

CLEAN UP AND FINISHING

Whether the terrazzo is the fine system with the 3-in. floating base or the thin-set system, it has a disadvantage to the builder that causes the author to rate it number 1 as a "nuisance factor." This is the slurry that comes from the grinding and honing. And if the terrazzo is Venetian, which uses larger marble chips, there is **more** slurry. If the terrazzo's matrix was composed of portland cement and water, the cement is now "dead" and will not further combine with the marble grindings during the grinding and honing work. The result is wheelbarrow after wheelbarrow of semiliquid "goop" that must be squeegeed to one side of the terrazzo floor and left until enough moisture (very slowly) evaporates from it and it may be shoveled into wheelbarrows and put into rubbish trucks. If someone could discover a use for this biproductal goop, he would make millions! If the matrix of a terrazzo topping is epoxy or acrylic, there will be little difference in the amount of slurry. The amount of slurry is determined by the amount of marble dust ground off.

After terrazzo is ground and honed, pure base material (i.e., portland cement, epoxy or acrylic, depending on the type of matrix) is troweled into the honed surface to grout in any minor holes left when small marble chips were pulled out by the grinding and honing processes. At a later date, this grout is polished off with a fine grinding stone. This final polishing process should be delayed as long as possible to protect the terrazzo. As little grinding water is needed in the polishing process, it can be delayed even after cabinetwork is installed.

One other caution to the superintendent; the flattened and honed marble chips are little sponges! A terrazzo floor must be protected from spillages of

oil, paint, or solvents. If there is a spill, get the terrazzo man in **immediately.**
He may be able to poultice out[1] some of the staining material.

Terrazzo is a troublesome product, but it is worth it. All beautiful things must be cared for and protected. A terrazzo floor with a well-chosen chip mixture and matrix color will beautify your building for as long as the building stands!

22–5   Poured Resilient Floor Surfaces

Another material that falls into the category of the wet trades is the poured resilient floor. It is a family of resilient surfaces whose base material and matrix change as required by the intended usage. There are acidproof and impactproof floors, and floors used for athletic games, for example. Generally, these systems do not have much of a moisture-nuisance factor. This is because, even though the surface is poured, it is generally not ground down; if it is one of the few products in this family that requires grinding, very little grinding is required.

22–6   Ceramic Tile

Regardless of methods of installation, there will never be a material that will replace ceramic tile for the walls and floors of toilet rooms and showers, even though sprayed ceramic-glazes are often tried by designers. Originally, ceramic tiles pieces (approximately 4 by 4 in.) were used with about 1 in. of cement mortar to apply them to walls. This was called the "mud system." After the ceramic tile walls were finished, a lean (one to three) mixture of cement and sand was used for floor fill and screeded out as a dry pack. After screeding out to match the level of the wall's ceramic tile base and curing, floor tiles (usually pasted onto paper) were laid onto a thin cement grout. The mud system was used for wall surfaces because the terra-cotta or concrete block partitions were not always adequately plumb. Later, as costs rose, and better cementitious tile adhesives were developed, tile mechanics applied a thin screed coat of cement plaster to the masonry of the partition, which was absolutely plumb, and the next day applied the ceramic tile squares directly to the screed coat with special ceramic cement. The floors were laid in the manner previously described.

Finally, costs becoming still greater, ceramic tile was laid directly onto gypsum board partitions with waterproof tile cement. Floors were laid onto troweled concrete surfaces, ¼ in. low, with the same cement. The resulting room looked the same. Thus, these days most of our ceramic tile is applied

---

[1] In medicine, a poultice is a mealy mixture put onto an inflamed area of one's body to draw out the poisons. In construction the process is similar, except that to remove oils a plaster-of-paris mixture is used.

to dry-wall partitions (the green-colored "moisture-resisting" type of gypsum board being used in toilet rooms) and the result is good. However, there is no substitute for ceramic tile on **masonry** for shower stalls. The author warns that showers formed with ceramic tile on gypsum board may cause problems that are never found when the tile is applied to masonry.

There is little nuisance factor with any type of ceramic tile system. Ceramic tile is expensive and its installers try to have as little waste as possible. Also, because the rooms are small, waste must be cleaned up before the rest of the room may be completed.

22–7   Wet Trades in General

We have defined the wet-trade materials as materials that produce wet refuse. A new building cannot have all its construction moisture removed (so that certain trades such as cabinetwork may be installed) until the wet trades have been completed and the wet refuse has been removed from the building. The big problem is the wet refuse. Far too many subcontractors "drag their feet" when it comes to getting rid of their refuse. It's always going to be done "tomorrow." Here again, the Construction Superintendent must be ever mindful of the construction motto: "Don't put anything off till tomorrow that can (somehow) be accomplished today." Following this motto may very well save your schedule and save your company's profit.

Chapter 23

Interior Finishes and Specialties: The Dry Trades

After we have listed the "wet" trades in a building's interior finishes, almost all the rest of the trades could be listed with the "dry" trades. When we discussed the wet trades we described the manner in which the systems should be installed and certain precautions to be observed. We shall follow this format with the dry trades. However, in addition, we may mention "lead-time" ordering for certain of these materials. Whereas all the wet trades have a disadvantage in that they produce troublesome wet debris that must be removed from the building, most of them have an **advantage** in that the materials for their manufacture are usually readily available locally. Terrazzo, which sometimes uses special marble chips, is one of the exceptions. However, many of the dry trades require special materials that must be ordered long before the system is to be installed in the building.

A partial listing of the dry trades would include the following:

Dry-wall partitions (8)

Hollow-metal doors and bucks (2)

Movable metal partitions (3)

Hardware (1)

Acoustic tile ceilings (5) to (7)

Factory-made resilient flooring (6)

Carpentry, millwork, and cabinetwork (4)

Wall coverings, including paint (5)

These dry trades are listed in the general order that they might be installed in a building, with the exception that carpentry work goes along during the entire building process, from the foundations up. The numbers *after* the trades are not "nuisance-factor" numbers, as they were with the wet trades; rather, they indicate the order in which the purchasing department of a construction company should start the purchasing or ordering procedures. For example, because hardware has so many different types of devices that must go into a building and because hardware companies are noted for needing a great deal of lead time to get the orders into "production," hardware is one of the first things a construction company arranges for after structural steel (another long-lead item) is "bought out."

The companies that manufacture hollow-metal doors and door bucks schedule their production lines months in advance; in addition, door bucks and doors cannot be manufactured until templates for their hinges, locks, and other hardware are available. Thus construction companies usually order hardware first and hollow metal second. The hardware manufacturer starts producing his lists of hardware (called hardware schedules) for the Architect's approval, and the hollow-metal manufacturer starts making his shop drawings for the Architect's approval. As soon as the hardware schedule is approved, the hardware company can send cardboard templates for the hollow-metal company to use in preparing the cutouts in metal doors and metal door bucks so that the hardware will fit. Movable metal partitions have the same problems. Therefore they are listed as (3).

Millwork and cabinetwork require shop drawings, special veneers, and careful manufacture. Therefore, it comes next on the list. Wall coverings are often specials, and it takes considerable time to get them into the production schedule of a company. Sometimes an Architect's interior designer specifies a specially patterned Belgian linen that must be woven to order. In such a rare case a (5) might be too high a number. Acoustic tile ceilings do not, usually, present a long-lead ordering problem. However, once in a while a designer will call for some special feature in the acoustic tile's suspension system. These special systems are not universally used and, therefore, must be made to order. For this reason we have put a variable (5) to (7) as an order factor.

Resilient tile is not really a problem. However, because there are thousands and thousands of patterns, most floor-covering manufacturers do not stock tile, and their installers stock only a small amount of certain patterns that they feel they can job out for small installations. However, whenever large floors (10,000 ft^2 or more) are required, the tile must be manufactured to order. Thus we rate resilient tile with a lead-time number of (7).

Finally comes dry-wall partitions (i.e., gypsum board on metal studs). This system is so universal that most of the materials are what we call in the trades "shelf items." That is, most suppliers have plenty of the materials on stock. The only problem might occur when an especially high partition is required that would need a stronger metal stud. These studs might not be in local stock, but they can usually be ordered for immediate shipment from the factory. Thus we put a lead-time number of (8) on dry-wall systems. Our discussions from this point on will deal with methods and cautions. We shall work in the order of the list.

23–1   Dry-Wall Partitions

The dry-wall partition, a partition of metal studs set in metal channel tracks with gypsum board attached to it with special drive screws, has almost supplanted the original plaster-on-masonry partition. It is less expensive and can be erected by mechanics who have less experience. One is tempted to say "less ability," but this is not true. It takes long experience before a bricklayer can erect an interior partition that is straight and plumb, and it takes a long-experienced plasterer to lay a true coat of plaster on this partition or wall. However, whereas almost any carpenter can set up a gypsum board partition, it takes a **good one** to get it up fast and true. And getting it up fast is his employer's first requirement.

It would seem almost impossible to put up a crooked dry-wall partition. All that is necessary is to snap a chalk line, fasten the base track to the floor with powder-actuated nails, and set the partition studs plumb. Then why are there many partitions so crooked that a snake might break his back following them? Lack of experience or lack of pride. But a dry-wall partition *can* be laid true to line with very little experience and just the will to do a good job.

Dry-wall partitions are usually more than just a layer of gypsum board on each side of metal studs. Very often there are two layers of gypsum on the corridor side and, when this is the case, the joints of the outer layer are offset or "staggered" 16 in. from the joints of the lower layer. This double layer (along with the single inner layer) is placed to give the partition assembly a longer fire rating. If acoustic levels are important, the two outer layers can be separated by a ½-in. spacer channel, and rock-wool or glass-fiber insulation can be placed between the inside and outside layers of gypsum board (i.e., between the metal studs). If door bucks are the conventional type, they must be set before partition installation begins, just as they would be set before a masonry partition was started. However, door bucks can be procured that come in two pieces (one for the corridor side and one for the room side) that mesh together. In such cases, double studs are set on each side of the opening, and the bucks are set after the partitions are built. This is a time saver when there is a rush job and bucks could not be ordered in time.

Dry-wall systems are also available for such important installations as elevator shafts. The basic gypsum board is two 1-in. layers of gypsum board, factory laminated into a 2-in.-thick board, 24 in. wide. The 2-in. panels are placed into 2-in. base and head channels, and 2-in. tees are placed between the 24-in.-wide boards. After this system is erected, additional layers of ½-in. gypsum board are applied on the outside of the shaft and also, if the fire rating requires, on the inside. This system is very durable and can resist the suction and pressures of fast-moving elevators. The system is more expensive than masonry for elevator shafts, but there is a saving in cleanup. As it is used more extensively, the price difference may become negligible. One of the first high-rise buildings to use this system was the Central National Bank Building in Cleveland, and the twin towers of the New York City Trade Center (second highest buildings in the world) followed shortly thereafter.

Perhaps this heading should read "Hollow-metal bucks and doors." Actually, it is the doors that are hollow but the bucks are always linked in the hollow-metal trade. What is important is that the bucks[1] are required far sooner than the doors. The bucks must be shown on shop drawings at the same time as the doors, and the whole assembly must be checked and approved by the architect. Thus shop drawings and the checking must be expedited. Once the bucks reach the job, the builder is safe. His Field Engineer has hopefully laid out all the partitions (whether they are masonry or dry wall), and the carpenters can now erect the bucks. It is important that the engineer's layout be flawless, and it is just as important that, once the bucks are set, shimmed up, and nailed to the floor by the carpenters, the carpenters adequately brace them to the ceiling so that they will remain plumb until the partition is built to hold them.

As noted in the introduction of this chapter, hollow-metal work is a long-lead item. There are not too many manufacturers of metal doors and bucks. It behooves the Project Superintendent to put pressure on his Purchasing Agent to make the original order, and it is important that he or the Project Manager keep tabs on the Architect to be sure that the shop drawings do not lay on some checker's desk for weeks. Be very aware of one thing. The manufacturing of door bucks is a very competitive trade and a lengthy operation. No hollow-metal manufacturer, computerized or not, will start a job until *all* the drawings are checked and approved.

23–3   Movable Metal Partitions

Office buildings are usually designed to show (only) partitioning for the elevators, toilet rooms, electric closets, and slop-sink rooms. These partitions comprise what is termed the **"core"** of the building. The other partitions of each floor are designed by the **tenant-change** designer. In tenant-change design, many of the corner offices and often all the exterior offices (i.e., those which have exterior views) are designed with permanent partitioning. However, if the tenant feels that there may be a change in his operation during the years of his tenancy, he may install metal, metal-and-glass, or "bank-screen" metal partitions. These can be fabricated so that they run from floor to acoustic ceiling; if sound transmission is a problem, lead-sheet baffles can be installed over the acoustic ceilings to the upper concrete slab (in line with the metal partitions) to give additional acoustic barriers.

The advantage of these partitions is that, if the operation of an office changes, the 4-ft modular sections may be removed and set up to form other office layouts. However, the partitions are more expensive and must be ordered well in advance if they are to be ready when the progress schedule calls for their installation. Like hollow-metal work, their manufacture de-

[1] Buck is a trade expression for prebuilt door frames.

pends upon hardware templates. Therefore, here is another reason that hardware must be ordered as soon as possible and that the correction and approval of hardware schedules be expedited.

Movable partitions, especially the lower bank-screen partitions, are *sometimes* made in wood. These are usually exotic laminates such as rosewood and are a part of millwork, which we shall discuss in Section 23–7.

23–4   Hardware

The manufacturers of hardware items (which include locksets, hinges, doorplates, escutcheons, door bumpers, door closers, and scores of other items) usually work on a yearly schedule. That is, they forecast what amount of certain items (such as cast doorstops) will be sold in a year and then they manufacture that amount plus a cushion (perhaps 10 percent) in addition. Therefore, if your construction company needs cast-bronze doorstops (often plated thereafter), they had best be included in the casting schedule of a hardware manufacturer or your company will wait 12 months for the next casting. Were hardware manufacturers to deviate from this policy, they would soon go out of business in a very competitive endeavor.

The hardware for a house can be picked off a hardware store's shelves. However, the keying systems, the configuration of locksets or latchsets, and even the butts (hinges) must conform to all the hardware in a large, commercial building. In addition, the keying system must be individual for each floor or tenant, have a master-key system for each floor or tenant, a master-key system for certain universal rooms, such as electric closets and slop-sink closets, and have a "grand-master" key system to cover the whole building so that, in an emergency, the Building Superintendent or his authorized assistant may enter any space. This keying schedule takes the brains of a genius and the patience of Job and takes lots of time.

Most certainly door bucks, doors, and movable metal partitions may be manufactured as soon as hardware schedules are approved and hardware templates are available. However, it is going to be many months before the hardware itself is ready for installation. Of course, it is the responsibility of a construction company's purchasing department to expedite this work and the approval of the architects. However, when a building is late in construction completion or exceeds the company's budget, it is usually blamed on the superintendent even when he documents his problems. Thus the Project Superintendent should watch the hardware situation to be sure "his building" is covered.

23–5   Acoustic Tile Ceilings

Acoustic tile ceilings are, generally, suspended from the slab above with No. 8 wire or with **pencil rods** (a slightly thicker metal rod). The most important attribute for a suspended ceiling (whether it be plaster on lath or acoustic tile on a metal suspension system) is that it be at the right elevation and *very* level. The grades are too critical to be set with a rod and a surveyor's level. Of course, 4-ft grade marks have been set by the Field Engineer onto the walls and columns of a floor. From that point on, however, the ceiling installer depends upon only **one** of these 4-ft marks and runs his ceiling grades himself. Until recently, a lather or ceiling-suspension contractor used a long hose with sight-glasses filled with water. He set his initial suspension grade from the Field Engineer's 4-ft mark, and then used this ceiling grade and his "water-level" to transfer grades to all the hangers for that ceiling. Regardless of new systems, there is no method more accurate than the water-level system. However, this system takes time.

In recent years, ceiling-suspension contractors are using a laser beam on a tripod, and setting the level of this laser beam most accurately from one of the Field Engineer's 4-ft marks. Thereafter, the laser rotates like a small lighthouse and throws a very thin beam onto a hanger rod or small target. Thus the lather suspends all his hanger rods and then marks them when the beam of the laser (which is leveled very much like a surveyor's instrument) hits the rod or a target. A suspended ceiling should be installed for the entire floor area where there may be relocation of partitions, just as we recommend for floor finish. If this is done, minor patching may be required, but the level and the module will be consistent.

The original suspended acoustic ceilings were constructed with inverted tee sections, which supported the acoustic tiles. Next came the *concealed-spline* system, which uses an inverted tee for main modules, but has the tee going into slits (or kerfs) at half-depth of module tiles, with smaller, independent, tee sections supporting the tiles between the main tees, which are hung from the floor above. Both systems use materials that are usually shelf items. However, there are a number of newer systems that use suspension angles and suspension tees, which supply special features such as air-handling slots for air-conditioning systems. These systems are "long-lead" systems and must be ordered well in advance.

23–6   Resilient Flooring

The first resilient flooring in the building business was linoleum. This came in 4- to 8-ft-wide rolls and was applied to the structural floor with linoleum paste. More durable linoleum floors were constructed with ¼-in. "Battleship Linoleum" which was actually used in battleships. Its advantage was durability. Its disadvantage was that it was usually not available in colors other than basic grey or basic brown.

Then the resilient flooring manufacturers came up with the 9- by 9-in. and 12- by 12-in. tiles, which were ⅛ in. thick. The first of these tile systems came in asphaltic materials that were stronger when the color selection was of the darker (more asphalt) tiles. Asphalt tiles have the advantage that the material breathes, and moisture that may enter between joints during floor washing can evaporate through the tile. Asphalt tile has an advantage that it is less expensive than other resilient tile. However, it has the disadvantage that it is not as hard as some other tiles such as vinyl-asbestos tile.

Vinyl tile is more expensive. It has the advantage that, for months after it has been installed, it will expand slightly and thus close the tiny joints. It is softer to walk upon. However, it does not breathe like asphalt tile. Therefore, if any water enters the (narrower) joints because of sloppy cleaners who allow water to lie on the surface for longer periods, the water must "travel" under the tile in order to dissipate itself. In this process it will destroy the adhesive properties of any linoleum pastes (which are often used for rubber or vinyl tile) unless the adhesive is of a waterproof nature. The Construction Superintendent should "watch" this.

Vinyl-asbestos tile is probably one of the most durable of the resilient floor materials. It is very tough and will withstand spike heels which the ladies sometimes wear far better than its forerunners. It has the advantage that it "breathes." It is more expensive than some of the other tiles, but it is worth it. It lasts longer and its finish is harder.

There are a number of other resilient flooring systems. There are tiles of ¼-in. cork, tiles made up of ⅜-in. waffle-system rubber or vinyls, and a whole salesman's suitcase of specialized flooring systems. It behooves the Project Superintendent to be sure that all tiles are installed in strict accordance with manufacturers' recommendations and with the cements recommended by the manufacturers.

Because resilient flooring must be installed before certain other trades can complete their work, it is subject to scratching and damage. Workmen carry abrasive dirt on their feet especially when entering a floor from a stair tower. Thus a good contractor will wish to cover the resilient tile wherever possible, but *always* for 50 ft (of foot-travel) from the stair towers.

23–7 Carpentry, Millwork, and Cabinetwork

Millwork and cabinetwork are installed by carpenters. On a construction project most General Contractors keep several good carpenters on the project from the very beginning. These carpenters will build the temporary protection barricades throughout the project, and they will install certain other architectural appurtenances that the General Contractor has not included, in addition to installing wood grounds for terrazzo base or grounds for wood base so that a plastering contractor's work meets wood base which will be installed later. Of course, if the General Contractor is building the forms for concrete foundations and for concrete floors with his own men, there will be a much larger carpentry force. Regardless, there should be a nucleus of "finish" carpenters to do the finer work.

Cabinetry and millwork are the product of much shop fabrication after large-scale shop drawings are drawn and approved. Carpentry for the field portion of this work will be done by special carpenters, who will set grounds for wall paneling and for the support of cabinets. After these grounds have been installed, a draftsman from the cabinet shop will field-measure and place the field-measured dimensions onto the shop drawings so that the millwork (stairs, wood doors, chair-rails, base, wainscoting, bank screens, wood panels) and cabinets will fit the grounds and spaces exactly.

The precautions that a Project Superintendent must take for millwork and cabinetwork are to be sure that shop drawings are made and approved as soon as possible and that, following this, grounds and back-plastering (when the latter is necessary) are installed long before the millwork and cabinetwork is to be installed. And, after millwork and cabinetwork are delivered to the project, they must be protected from moisture and physical damage.

23–8   Painting and Wall Covering

There are a number of old sayings and rhymes about the building business that come to the ultimate conclusion that the painter must take over all the faults and sins of trades that came before. This is somewhat true. Thus a good General Contractor or Project Management organization will be sure to provide enough "temporary light" so that the painter can see all the dents and blemishes in the wall he is preparing for paint or wall covering. The plaster or gypsum board must be prime-painted, and then "spackle" (mechanics usually call this "sparkle") must be laid into the dents and low spots and covered with more prime paint or primer sealer before second and third coats of paint are applied to the wall, partition, or ceiling. **Good** preparation is **absolutely required** before finish paint or wall covering is applied.

Wall coverings range from wallpaper, to synthetics, to linen backed with paper. All these coverings are at first more expensive. However, in the long run they may be more economical. Vinyl covering is always recommended for public corridors where dirty hands soon ruin painted surfaces. The vinyl coverings can be washed for many years. Thus, if the owner is to spend extra monies for wall coverings that he will view for many years, it is even more important that the walls under these more expensive wall coverings have no blemishes in their surface.

Vinyl wall coverings and wallpapers are often no more than swatches in a salesman's book ready to be manufactured when enough of the material is ordered to warrant a "run" in the factory. Thus, if the client has chosen a paper or fabric wall covering, it should be ordered as soon as possible. Unless it is a small job where a few shelf-held rolls will suffice, wall fabrics are long-lead items and should be ordered well in advance of the time when they are scheduled to be applied.

23–9   Attic Stock

When a large building is erected, a number of special items are used which, at a later date, may not be readily available or will not be available at all. Therefore, in most commercial buildings, the owner orders a supply of matching items so that he will have these supplies available when alterations or repairs are made.

For instance, patterns of acoustic ceiling tile, resilient floor tile, ceramic tile, and wall coverings are just a few of the building finishes that will undoubtedly be discontinued in a few years for new patterns. Thus the owner of the building should have a certain amount of these materials available for repairs or future alterations. Most certainly, metal door bucks *could* be reproduced to match the door bucks originally specified. However, if the hollow-metal fabricator has discarded the special dies he made for an architect's special details, it will be costly indeed to make door bucks to match "building standard." Thus, in the old days, an owner ordered extra amounts of these special materials, which he stored in the attic of his building. Now, when the top floor of a building is usually the most desirable, extra stock materials are stored in cellar storage rooms until they are required. They are still termed "attic stock," however.

A good builder will suggest certain percentages of special materials for the owner to buy as attic stock. In addition to a number of right- and left-hand door bucks, the owner would be well advised to buy extra doors and locksets. We have mentioned wall coverings as an item that should be purchased for attic stock. However, paint is an item that should *not* be purchased for attic stock because the unopened paint will never match paint that has been on a wall for a year or more.

23–10 Keeping the Dry Trades Dry

Any of the materials mentioned in this chapter will be harmed by moisture. For instance, it does not make much sense for a specification to require that the hollow-metal manufacturer bonderize[2] his door bucks and doors and apply special prime paint or finish paint only to have this material lie around in a damp building. When it comes to gypsum board and millwork, dampness can ruin these products. None of the materials mentioned in this section should be brought onto a project or installed until they can be placed in an area that can be kept dry and without excess humidity. Furthermore, for wooden doors and other millwork, these materials should not be brought into a space with less than 50° F temperature in addition to a space where humidity is properly low.

Remember, the finishes of a building are what make a building a good-looking finished product. Protect the finishes and you protect your company's reputation and profit.

---

[2] Borderizing is a process wherein metals are chemically etched so that paint will bond to the metal more effectively.

Chapter 24

The Mechanical/Electrical Trades: General

When one mentions the mechanical/electrical trades, one thinks first of plumbing, steamfitting, forced-air heating and cooling, and the wiring of electrical devices. However, there are many more trades involved in this grouping, and the purpose of this chapter will be threefold: (1) to list some of the trades, (2) to discuss when these trades should be brought onto the project, and (3) to list how some of these trades are usually "lumped together" in single contracts. In Chapter 25 we shall discuss the coordination that must be accomplished so that the mechanical/electrical trades can install all their work without conflicting with each other and with the architectural–structural aspects of the project.

For unknown reasons, but probably because trade delegates for the steamfitters pushed harder once, the plumbing trades' work (depending on local "trade" agreements) is usually limited to the supply of hot and cold water, the installation of lavatories, toilets, showers, and hot-water heaters, water supply to steam boilers, the piping of illuminating gas (now used for cooking and heating fuel), floor and roof drains, and the heavier drain lines that carry the wastewater from these units to the storm and sanitary sewers in the street.

The sheet-metal workers (often called "tin knockers" because they install the duct elements by knocking the joints together with a hammer and a hand anvil) take care of the ductwork throughout a project, and they rig and install the fans (or air-handling units) that move air through the building. The steamfitters pipe the internal drain pans of the air-handling units to the external floor drains, but plumbers install these floor drains and the heavy drain lines from them. And, understandably, although electric motors for air handlers are usually supplied by the contractor who supplied the units, the electricians wire the motors and circuit breakers into the main, electrical system.

Besides the installation of steam boilers and the piping from these boilers, steamfitters also handle piping for chilled water, gases such as oxygen, hydrogen, nitrogen, helium, or even compressed air (which would be used in a laboratory or a hospital). In addition, there is smaller piping for pneumatic thermostats and thermostatic controls. Whereas this has, until recently, been handled by steamfitters, there are now specialized unions cropping up in certain areas that claim this work. And, at this point, we might wander from the basic subject of this chapter to discuss the claims of certain union officials such as shop stewards and union delegates.

203

The
Mechanical/
Electrical
Trades:
General

On almost every project there is an "odd ball" item that is so rarely used that more than one union will claim it as soon as the material comes onto the project, even though the General Contractor has bought the work from a specialty contractor. For example, on one of the author's projects, we used waterpipe for concrete pipe screeds. The steamfitter shop steward tried to claim it. We handled this matter as we advise a Project Superintendent to handle **any** jurisdictional problem: (1) we asked the steamfitter's delegate to show us where pipe screeds were covered in his book (he claimed it was "special piping"); (2) we looked in the International book and found that screeds were awarded to both the carpenters and the cement finishers; (3) we called the stewards of the cement finishers (the cement finishers were setting the screeds with the help of the cement laborers) and the carpenters to a meeting with the steamfitter steward and discussed the matter. The carpenter steward was satisfied that his union was erecting the forms for the concrete and was setting the grade nails at the edges of the forms. Thereafter the Project Superintendent (at that meeting) ruled for the cement finishers and their laborers. In addition, he advised the steamfitter steward that he could confer with his delegate and come back to the Project Superintendent for further discussion, but that otherwise the discussion was closed.

Basically, most jurisdictional problems can be solved by reading the contract books of each trade; when two unions may have been granted something that creates a borderline case, the Project Superintendent should look into the International awards book. There are certain trades that have problems that should be solved **long before** the material comes on the project. An example is corrugated asbestos siding. Originally, corrugated siding was sheet metal and was therefore claimed by and awarded to the sheet-metal trade. However, when a manufacturer produced a siding material in corrugated asbestos board that could be cut with a carpenter's saw, it was natural that the carpenters would claim this material, and it was natural that the tin knockers would make a counterclaim.

Disputes lose time for the trades and the contractor. The corrugated asbestos siding example is one of the few disputes we encounter in the architectural trades. However, with new heating and ventilating techniques, disputes can be numerous. Therefore, when a General Contractor sees a controversial item in the contract items, he should call the delegates together **long before** the material arrives and make sure the jurisdiction is decided as soon as possible.

Now, let us return to the pipes, ducts, and wires.

Electrical work is usually done by electricians without any problems. However, there are sometimes two electrical unions in a territory. For example, in most areas the electricians union will do all the wiring for motors, lighting, and low-voltage bells and signals, and will allow telephone company men to install the telephone wires and electricians for bank vault alarm companies to run the burglar-alarm wiring. In other localities the electrical unions will not allow this, and there will be a dispute. The possibility of such disputes must be considered by the General Contractor and the Project Superintendent.

24–1   Temporary Services

As soon as a construction project commences, there is need for temporary service to be supplied by the electricians, the plumbers, and sometimes by the steamfitters. By the time the project commences, the General Contractor or the Project Management organization will have contracted for or bought-out the mechanical/electrical trades. In these subcontracts, and in the specifications there will be notes that "temporary services" shall be provided by the appropriate trades.

For instance, as soon as field offices and workmen's shanties are erected, there will be a need for temporary lights and power to these shanties. And as a building is erected, there must be temporary lights in the stair towers and other areas in the building.

Whereas chemical toilets may be used by the workmen, it is a much more satisfactory arrangement if a temporary toilet room can be erected in the basement area of the new building (on stilts so that concrete floors can be poured when scheduled) to house a number of flush toilets and lavatories for the men. Such a convenience can be erected soon after foundation walls are built and the plumber has brought his permanent sewer line into the basement. Chemical toilets will suffice and cover local, union, and OSHA codes. However, the workmen appreciate a more substantial facility and tend to keep it cleaner. And as the height of a building increases, temporary toilet facilities may be placed into permanent toilet-room areas so that the workmen may be comfortably served. We emphasize how much happier a project can be achieved by providing clean and comfortable temporary toilet and washing facilities. When a workman isn't comfortable, he finds other places that will make all other workmen uncomfortable. Temporary washrooms with flush toilets cost a few more pennies, but they are worth it.

As soon as contractors and subcontractors come onto a project, they will have telephone service installed in their field offices. However, to assist all workmen and to keep workmen from making unauthorized use of company telephones, the General Contractor should arrange for one or more pay phones to be placed in areas where full-time staff can observe them and insure that they are not abused.

24–2   Temporary Heat

There are two kinds of temporary heat. There is the temporary heat that is provided to keep new concrete and new masonry warm and there is the temporary heat that keeps the workmen's shanties warm. In most cases the heaters use oil or propane. Either of these fuels can bring disaster to a project if local fire department regulations and safety standards are not observed. The fuel must be carefully placed (i.e., there should not be too many drums or tanks stored in one area, and the fuel-supply should not be too close to the heaters), and the heating units, heated areas, and storage areas should be well policed to avoid hazards.

24–3   Permanent Services

Eventually, all mechanical/electrical services will be installed in the building or the entire project. However, the plumber must bring his illuminating or heating gas, his sanitary sewer, and his storm sewer lines into the building. The steamfitter must bring his oil lines into the building or, if the heat of the building is to come from city or project-manufactured steam, he must bring steam lines into the building. The electrician must bring supply conduits into the building, and must install conduit to be used by the telephone installer. All these services must, usually, come through the foundation wall of the building or sometimes, in the case of storm or sanitary sewers, under the peripheral footings. Therefore, as soon as one face of the foundation form work is installed, sleeves must be installed so that there will be holes for piping and conduit after the peripheral foundation walls are poured and stripped. If the designers feel that there may be a water-penetration problem, they may require that these sleeves have a ring welded to the sleeve to act as a water stop halfway through the concrete wall.

24–4   Temporary Use of Permanent Lines

In Section 24–1 we mentioned temporary washroom–toilet facilities. If the site is large and the sanitary sewer for a project has to come for some distance, the General Contractor may decide that he will place this temporary washroom–toilet facility outdoors to one side of the new, main sewer

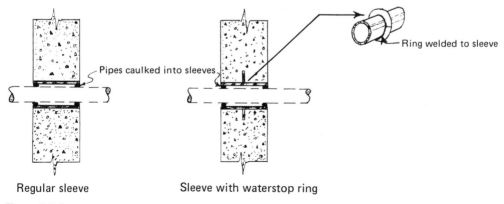

Regular sleeve   Sleeve with waterstop ring

Pipes caulked into sleeves

Ring welded to sleeve

Figure 24–1.

line, and bring the waste into this line by a temporary tee that can later be plugged. However, if project space is tight, he may ask the plumber to bring his permanent sewer line into the building, and will place his temporary facility between column lines and off the ground on stilts so that the permanent building and floor slabs may be built.

206

The
Mechanical/
Electrical
Trades:
General

In addition to sanitary sewage, there is always groundwater or rainwater being pumped from basement or cellar areas during the first part of the construction period. If the plumber has brought his storm sewer line into an outside manhole and from there into the building, the General Contractor or the Excavation Contractor may use these lines and pump groundwater into them. Otherwise, this water must be pumped past the curb and must travel down the street to the nearest curb catch basin. This might cause inconvenience to the builders or, worse yet, the neighbors. Thus, if a storm sewer line is called for, get it in as soon as possible. As we suggested in Chapter 8, plan for a dry project!

Temporary electrical power is seldom a problem. As soon as the Electrical Contractor comes onto a project, he calls for a tap from the local power company. As soon as this power is brought from the pole or from a street conduit, the electrical contractor brings the line into a temporary plywood power-center box where he has cutout switches for temporary light and temporary power. As the building progresses, he will move this power center inside. However, in most cases he will *not* take temporary power from the main switchgear room (even though feeders are pulled to the street) until most of the entire main switchgear is installed. This gives him more room for his permanent work. Nevertheless, this contractor and contractors for the mechanical trades must be pushed to place their sleeves into the forms so that structural progress will not be delayed.

24–5   Excavation for Mechanical/Electrical Services

Most sensible contract documents call for ditches and excavation for mechanical or electrical lines to be performed by the particular trade that will use them or, at least, by their own excavating contractor. Often, if this is the case, the mechanical or electrical contractor will pay the excavation contractor to dig these ditches while he has a piece of excavation machinery on the project. However, regardless of contractual obligations for the main ditching or excavation, the bottom 12 in. of the ditch should be hand excavated and backfilled by the contractor who will place his lines into the ditch. For instance, if this ditch is for a plumber's drain line, the bottom of the ditch must be accurately excavated so that the inside bottom of the pipe (or "invert") is at the correct level. After this line is placed accurately in the bottom of the ditch, the plumber should be responsible for backfilling the ditch with good granular material (especially no large stones) so that the sewer line is protected before other, less granular material is placed into the trench. If the designer or the construction organization sees that this duty is placed upon the contractor utilizing the ditch, there is no chance that the mechanical or electrical contractor can claim damage to his line.

24–6   Temporary Placement of Mechanical or Electrical Equipment

We have all heard of people who built a boat in their basement, only to find that the final product did not quite clear the openings they had intended to use to route it to the outside world. There is a very good chance that a similar problem can come to the builder of a large structure.

All buildings and projects have elevator machinery, switchgear, chilling compressors, and tanks (to name a few) that could not pass through the doorway to the rooms in which they will remain. Of course, some might be disassembled and moved into the room after it is built, but usually it is easier and more sensible to lift or rig the piece of equipment into place as soon as the area can receive it. For instance, even though an elevator contractor could lift his machinery 16 floors from the street level to the elevator machine room beams using chain falls, he usually pays the steel erector to lift his equipment with a crane as the steel for that level is finished. If the building's frame is reinforced concrete, the concrete-lifting crane is used as soon as the concrete deck, its shores, and load-spreading planking can hold the load. As soon as possible, machinery is moved to its final location so that the room may be built around it. The contractor who is responsible for the final installation of the particular piece of machinery usually makes the arrangement with the contractor who owns the crane.

24–7   Grouping or "Lumping" of Specialty Contracts

There are between 50 and 100 specialty contracts or trades in a construction project. However, many of these trades serve the heating, ventilation, and air-conditioning (HVAC) aspects of the project, the plumbing aspects of the project, or the electrical aspects of the project. Thus, although a General Contractor or Project Management organization might make prime contracts with each different trade, it usually groups the mechanical/electrical trades into three to four main contracts.

For example, the HVAC contract usually embraces all the steam or tempered-water piping, the sheet-metal trades (including fans and air handlers), fan coils or radiation units, electric and pneumatic thermostatic control systems, and all the mechanical/electrical machinery or devices aligned with HVAC system. All this equipment is installed by the HVAC contractor or subcontractors under contract to the HVAC contractor. Of course, the wiring of electric motors and breakers is usually the duty of the **electrical** contractor. However, if there are electrical connections in the thermostatic systems, these are usually covered by an HVAC subcontractor.

Plumbing fixtures, drains, pipes, and sewer connections from the building to the street sewers may be accomplished by more than one contractor. However, this work is usually provided by a plumbing contractor and one or two subcontractors under his contract.

In many areas, fire-protection systems, including the sprinkler system, are claimed by steamfitters. In some areas this work is claimed by plumbers. Regardless, the work in this subsection is usually done by another "prime" contractor or subcontractor to the General Contractor.

However, because all these air-conditioning and pipe trades must fit into adjacent areas and ceiling spaces, it is a happy General Contractor or Project Manager who can find one contractor who will take on the responsibility for HVAC, plumbing, and sprinkler work, even if this contractor does the work under a **joint-venture** arrangement between an HVAC and a plumbing outfit. As we shall learn in Chapter 25, much coordination is required between these trades. It is often worth a few more dollars to make one contractor responsible for all this coordination and cooperation.

Often projects call for many different electrical items that are not normally covered by the contractor that handles the electrical power supply and lighting. If some of these specialist-electric contracts can be economically lumped, there is less chance of division of responsibility.

Regardless of the minor differences in the grouping of the mechanical/electrical work of a project, there must be coordination; this is the subject of Chapter 25.

Chapter 25

The Mechanical/Electrical
Trades: Coordination

The machinery, piping, and ductwork that provide the environmental conditions in a building or factory are a formidable maze of material that must be designed, coordinated, and installed to fit into the structure so that they serve the purpose and do not detract from the architectural and structural design.

As the architectural and structural drawings are being produced, preliminary, "progress" prints are sent to the mechanical/electrical designers so that they may advise the architect as to how much space is required in machine rooms for the large pieces of machinery and equipment, and how much space must be provided in wall chases and ceilings for ducts, pipes, conduits, and lighting fixtures. Hopefully, the structure and the mechanical/electrical components are coordinated in the design stage so that everything fits. However, even when this design coordination is carefully accomplished, the contractors who supply and install the machinery, ducts, pipes, and conduits must carefully coordinate *their* work so that everything fits.

Years ago, when material was less expensive, Architects allowed more space between suspended ceilings and the structural beams above and more floor-to-floor space in mechanical floors. However, if an Architect cuts the floor-to-floor distance from 14 ft to 12 ft 6 in., he saves 1 ft 6 in. of column steel for every story; in a 20-story building this 30 ft of column steel will serve over two more stories. Of course, the column steel below would have to be heavier, but this is the general theory of floor-to-floor allowances. Even if no more stories are added, less steel (or concrete in a structural concrete building) is required.

In the coordination of the shop drawings for mechanical/electrical work, the ductwork and sprinkler work usually require a certain precedence. The ducts are the largest things that must pass through between the suspended ceiling and the bottom of beams and girders, and there is a requirement that sprinkler piping be kept level, continuously pitched, or at least have no dips or sudden rises. Because of the duct size, the usual shop-drawing coordination procedure (which all contract documents should require) is the preparation of large-scale shop drawings of these ducts. Because these drawings must be prepared in a large scale, they are usually called "bed sheets" and **210**

The
Mechanical/
Electrical
Trades:
Coordination
they are often close to the size of a bed sheet.

The sheet-metal contractor makes up the duct shop drawings and sends them to the printer for sepia transparencies.[1] A copy of these transparencies is given to each mechanical contractor and to the electrical contractor. Then each contractor plots his equipment (pipes, conduit, lighting fixtures, etc.) onto prints of these transparencies and brings these marked-up prints to a coordination meeting of the trades involved. Even if there is sufficient space in the areas involved, there is usually more than one contractor wishing to run his lines where another has set his. And all ducts and pipes must allow enough space for lighting fixtures when these are recessed into a ceiling.

The first intertrade coordination meeting usually proves that many lines have to move, and often the shape of portions of ducts must be altered while still providing the cubic feet per minute (C.F.M.) requirements of the air-conditioning design. Prints are "blue-penciled," and each trade's representative takes these back to his shop where he makes new prints of the duct sepias and works to alter the location of his installation to fit the requirements of others. Often these trade coordination meetings come to an impasse, and the Architect and his mechanical/electrical designers must sit down with the trades and help with the coordination.

Finally, when all trades are agreed that they can place the ducts, pipes, conduits, lighting fixtures, and other equipment into the space, all the equipment is drawn onto one transparency and submitted to the designers for approval. After approval stamps have been placed upon this, transparency prints are made so that all contractors and designers have prints from which to work.

When a contractor for a mechanical trade obtains copies of approved prints of these intertrade coordination drawings, he can proceed to make his own detailed shop drawings, which, when approved by the designer, will allow the contractor to fabricate his material and later install it. When the electrical contractor receives copies of the coordination drawings, he can order his equipment and start installing his conduits. Hopefully, coordination drawings will be made and approved in time for sleeves, inserts, and hangers, which must be placed onto the forms for concrete so that the desired hole or support will be in the concrete, and electrician's conduits may be installed in time to allow the structural contractor's work to proceed on schedule.

Of course, there are many areas in a building where there are not too many ducts and where there will be no problem in installing piping, ducts, and lighting fixtures. In such areas the design drawings are sufficient for the installation of sleeves, inserts, and conduit, and a slight "move" will be made by one trade or another as the material is set onto the concrete forms.

[1] Sepia transparencies are brown-line transparencies made from the original tracings by the blueprinter.

25–1   Field Implementation of the Master
            Progress Schedule

211

Field
Implementation
of the Master
Progress
Schedule

In Chapter 7 we discussed the master schedule for the entire project. Quite often this schedule was given to bidding subcontractors so that they would be aware of what would be required of them before they bid the project. Regardless, as soon as a contract is signed, **each** contractor should receive a copy of the master schedule so that he can order materials to be available when required and can bring the right amount of manpower onto the project when required.

In addition, the General Contractor or Project Management organization must hold weekly meetings with the trades to discuss the manner in which the schedule is being followed, and to allow the Project Superintendent to inform all concerned as to what they must do and when they must do it to meet the company's schedule or to bring the project back into line when conditions have caused delays. For example, the person conducting these meetings (usually the Project Manager or Project Superintendent) should advise the attendees at each weekly progress meeting where his company feels the project actually lies on the master schedule. Next he will outline any special action or increase of production that his company will require of the other contractors so that overall production will return to the master schedule.

Usually, the problem that confronts the Project Manager and Project Superintendent during the beginning of a project is that the mechanical trades are not installing their sleeves and inserts or the electricians are not installing their conduit in time to allow the concrete to be poured on schedule. If the Project Superintendent allows the mechanical/electrical contractors to run his project so that it fits their individual problems, he will soon lose control of the progress of the entire project. The Project Superintendent must watch progress carefully, and must require the contractors that work under his direction to have enough manpower to complete their work for each operation on time.

If, at a meeting on March 2 he advises all contractors that the concrete for a certain floor or a portion of that floor **would be** poured on March 16 in accordance with the master schedule, he should inform them that their portion of the work should be finished a day early in case of inclement weather in that next 2-week period. At the next regular meeting (on March 9) he should assay the progress in the past week and advise each contractor what he must do so that the pour schedule will be met. He may advise or demand that one or more contractors put more men on the project or use overtime (at their own expense) so that they will be ready on March 16. He may even use the old-time threat that, if their men are still working on the floor on March 16, "they will get concrete in their shoes." However, especially at the first part of a project, he must get the message across that *he* is running the project and that the project *will* go according to schedule.

Actually, because contractors and subcontractors make money if a project moves on schedule and lose money when a project is allowed to drag, the Project Superintendent should have an easy "sales-job" to convince the subcontractors or specialty contractors to complete their work on time. Of course, to do this, he may first have to approach the Management people for some of these contractors. However, once he has let a subcontractor's people know that he expects to do everything in *his* power to cooperate with them, and he expects them to do everything in *their* power to cooperate with him, his problems should be lessened. Regardless, he **must** show that he means business by keeping the project on schedule in the very beginning or he will never get it back onto schedule without much extra effort on everyone's part.

True, it is much better to *lead* certain people or contractors than to *push* them. Also, on every day of the project, he must assay the progress of everyone involved in that portion of the work and comment when he feels that one contractor is not doing as much as he should be doing. The Project Superintendent must be coordinating the project *every* day, not merely on every meeting day. For instance, he cannot expect the mechanical contractors to install their sleeves and inserts or the electrical contractor to install his conduit on schedule if the forms for the deck are not up to schedule. And, if he waits 4 days to discover that the form contractor is 2 days behind, or waits 4 days (until the next formal meeting) to tell this contractor that he is 2 days behind, the project will be 2 days behind and he cannot expect other contractors to put on extra manpower to solve a problem caused by the negligence of another contractor. The Project Superintendent must tell the delaying contractor immediately. He may even wish to confirm his warning in writing or by wire if the contractor seems reticent to comply.

However, bear one important rule in mind. A subcontractor who is making money is a happy subcontractor. A Project Superintendent can rely on cooperation from a happy subcontractor, especially if that superintendent is the prime mover in making good working conditions for all. Thus the superintendent should make every effort to see that areas are dry as soon as they can be kept dry, that areas are warm as soon as they can be kept warm, and that there is sufficient "temporary light" so that each workman can work efficiently. He should listen to a subcontractor's problems and try to help solve them. If a subcontractor's tardiness or poor workmanship is causing problems for others, he must take any steps necessary or be as tough as required to bring this subcontractor back into line. He may have to push one subcontractor so that he can achieve good working conditions for all and so that he may more easily lead the rest of the subcontractors.

25–2   Expediting Deliveries, Another Help to the Progress Schedule

A second major problem is that later in the schedule, equipment and machinery may not arrive from manufacturers on time. This problem is often difficult to deal with; therefore, the General Contractor or Construction Management organization should require assurances from contractors that long-lead equipment *has* been ordered in time. In addition, these contractors should be required to keep tabs on the progress of this equipment in the manufacturer's assembly and delivery schedules and report on same, when requested, at progress meetings.

Most contractors try to be sure that the equipment that they are to supply and install comes to the project on schedule. After all, they stand to lose more than others if it does not. However, sometimes these contractors are small organizations and do not have as much influence as a large General Contractor or, in some cases, the owner. For this reason most subcontractual documents have a clause which states that the General Contractor may require that a subcontractor (when requested) will provide the General Contractor with all pertinent information on orders, including order numbers, and that the General Contractor may expedite delayed orders and backcharge the subcontractor for the wages, travel expenses, and living expenses of their expediter. These expenses are often charged. However, in most cases, a few telephone calls from a large General Contractor or from the representative of the owner (who may currently be using a considerable amount of machinery or equipment manufactured by that supplier) will be very persuasive. Thus, unless the General Contractor has to spend a considerable amount of money on expediting, backcharges are seldom billed. The gain *to all* is worth more than the expense of putting through a minor bill.

25–3   Who Should Attend Progress Meetings?

Regardless of the contractual agreement between an owner and a General Contractor or between an owner and a Construction Management organization, one must remember that there are two separate teams on every construction project. There is the Owner–Architect team and there is the General Contractor (or Project Manager)–Subcontractor team. The Architect (who is a legal representative of the Owner) should meet with the General Contractor or Project Management organization on a regular schedule to insure that the contractors are working within the time frame of the master schedule; at this time the Project Manager may need to tell the Architect when the contractors need more information or more speedy reviews. The Architect may need to advise the General Contractor that shop drawings or material samples are not being correctly submitted. The owner may (and often does) attend these meetings.

However, before the General Contractor can be prepared to bring all necessary information to his regular meeting with the Architect and the Owner, he must meet with his subcontractors. These are the meetings we have discussed from one aspect in Section 25–1. At the General Contractor–Subcontractor meetings the discussions will be two-way discussions. As noted in Section 25–1, the Project Superintendent will be telling the representatives of the subcontractors where their companies are slowing the construction. Also, he will listen to complaints or suggestions from these

214

The
Mechanical/
Electrical
Trades:
Coordination

subcontractors as to how better cooperation from others will help them and the project in general. If the problems discussed are caused by the General Contractor or other subcontractors, the problems will be solved by this meeting. But if a subcontractor has submitted shop drawings or samples (through the General Contractor) to the Architect, and thus to his mechanical/electrical consultants, that have not been returned within due time, this information will be noted by the Project Superintendent and the subject will be one for the next Architect/Owner–General Contractor meeting.

The fact that there are two distinct teams is a fact that **no one** should forget or act contrary to. The Architect may meet with the General Contractor and may give orders that may, eventually, involve a subcontractor. However, and Architect or the Architect's representative may *NOT* give direct orders to a subcontractor, and the Owner, the Architect, or the Architect's representative *should not* attend the General Contractor–Subcontractor meetings. If the General Contractor (or Project Management organization) feels that a subcontractor has a particular problem that he can explain more accurately to the Architect, he will call a special meeting or he will request permission for the subcontractor to attend a portion of the Architect/Owner–General Contractor meeting. If, then, a subcontractor attends this meeting, his problem will be brought before the meeting first. After discussion of the subcontractor's problem, the subcontractor will leave the meeting and the regular meeting will continue.

There are exceptions to this rule. In fact, there are far **too many!** However, the legal entanglements caused by acting contrary to this general rule can be involved. Far too many large companies who have had other plants or offices designed by other Architects and built by other contractors have representatives who feel that they know more than their Architect, and they refuse to listen to his advice when he warns against "sitting-in" on General Contractor–Subcontractor meetings. These representatives feel that they get a better "grasp" on the project by attending these meetings and requiring that the Architect or his representative attend with them. Sometimes they **do** get a better grasp on the project; but, more probably, they are forced to grasp responsibilities of the contractors (that should remain with the contractors) because they "were a party" to the meetings.

We must mention that some governmental organizations require that the Architect and representatives of the mechanical/electrical designers meet regularly with the government's representatives, the prime contractors, and their subcontractors. However, many government contracts have clauses an Architect would find hard to set into a nongovernment contract. These clauses make it harder for a contractor to shift responsibility. However, the clauses do not solve all the problems. The author has attended many of these meetings, and he feels that they would be shorter and more productive if attendance were limited to the basic concept. If the contracts with a government agency require joint meetings, then these joint meetings must (regrettably) be held. On other types of contracts, however, the joint meeting should be avoided.

Chapter 26

Vertical Transportation

26–1   Definition

The term "vertical transportation" *could* include the movement of persons up and down the stairs of a building, but in the building trades **vertical transportation** means motorized transportation that takes a person from floor to floor without straining his physical systems or otherwise speeding his blood corpuscles. Thus vertical transportation includes elevators (moved by electric motors and hydraulic pistons) and electric motorized stairs.

The original elevators were hydraulically operated. Water was pumped to a pressure-tank on the roof of a building and was from there piped to a cylinder-piston assembly set into a hole drilled down into the earth and rock below a building's basement. Obviously, inasmuch as the piston had to push the elevator cab a certain number of stories, the cylinder-piston assembly had to be the same depth *below* the first lifting elevation as the distance between the lowest and the highest elevation of the cab's travel. The water-activated cabs were the original system and are still in use in many cities. The system provides vertical transportation of excellent speed.

However, as buildings became higher, and because there was a financial restriction to the depth a cylinder could be lowered with a piston fabricated with larger diameter to resist lateral bending where the distance between the top of the cylinder and the highest floor, vertical transportation turned from hydraulic elevators to elevators raised and lowered by means of cables and electric motors. Nevertheless, the use of hydraulic elevators did not die. Whereas the high lifts were served by electric motor–cable systems, the hydraulic elevator, which now used oil instead of water, took over the shorter lifts, especially those which required greater cab loads. The oil is not held in roof tanks but, rather, is pushed into the cylinder by an electrically powered pump. The vertical speed of these cabs is not great, but, when required, the system can be designed to handle great loads. For instance, in one of the author's buildings (a major bank), the security (money) trucks drive onto the elevator and are lowered one floor to the vault level. In other instances oil-powered hydraulic elevators are used when only three or four floors are to be served. The cabs move fast enough for the short travel required and a slower speed is no problem. However, within this range, the oil-hydraulic elevator will move a considerable number of passengers at a lower installation cost than will the cable–electric system.

Motorized stairs are useful for one- and two-story transportation or (as is the case in department stores) where people wish to ride for one or two stories and then board later for other floors.

26–2   The Universal Problems of All Vertical Transportation Systems

Whether it be an elevator or a motorized stair, the unit is made for the particular building, and the equipment is **long-lead** equipment. In all cases, the holes in the structure's floor slabs cannot be formed until initial shop drawings for the elevators or motorized stairs have been checked and approved by the Architect and his structural designer. Once the shop drawings for a motorized stair have been processed, the manufacturer can provide the inserts that will be cast into the concrete to support the machinery systems.

The installation system of a high-rise elevator is somewhat more complicated. Primarily, the holes in the floors must be most accurately poured so that the beams that will support the elevator rails, when set, will be plumb without the use of many shims. Most elevator companies like to keep shims to a minimum. Thus, after the approval of initial shop drawings, greater care must be taken in layout.

26–3   The Time Required to Complete a Conventional (Electrocable) Elevator

The operation of installing an elevator falls, in sequence, as follows:

1. Preparation and approval of shop drawings.
2. Measurements of the finished elevator shaft so that rail brackets of the correct length may be ordered.
3. Installation of rails to guide the elevator cab and the counterweights.
4. Plumbing adjustments to all these rails.
5. Installation and "roping" (i.e., running the lifting cables) of the car sling and counterweights.
6. Electrical connections and adjustments to the hoisting motors and their electrical controllers.
7. Cab installation.
8. Minute adjustments to the system.

Note: The sequence of raising the motors and controllers to the elevator-room level (as discussed in Chapter 24) varies with the height of the building.

When one considers that it may take two to three months to prepare shop drawings and obtain their final approval, and when one considers that some elevator companies will not fabricate all machinery until final approval of shop drawings, one realizes that elevators are, indeed, long-lead items. Thus the contract for them should be closed early in the project's negotiations, and shop drawings should be expedited.

26–4   A Warning—Do Not Use Elevator Shaftways for Temporary Material Hoists

Section 26–3 has given a brief outline of the many operations that must be completed before elevator constructors will complete an elevator. Under the **best** conditions, this time period ranges from 9 to 12 months. If a project needs temporary material elevators ("hod hoists"), they should be constructed on the *outside* of the building. When this is not possible, a separate opening should be left on each floor at some distance from the regular elevator installation. But **never** under any conditions use the shaftway for one of the elevators for a temporary hoist. The installation of the elevators is an operation that should be helped and expedited from the very beginning of a construction project. If one of the shafts for an elevator is used for a hod hoist, the elevator constructors cannot (**and will not**) work in the adjacent shaft.

One of the important aims of a construction program is to get one or more elevators finished as soon as possible so that they can be used as man lifts and, secondly, as material hoists. When one considers that a hod-hoist tower that will serve 20 floors can be assembled (or disassembled) in 2 weeks, and that the installation of an elevator will take 6 to 9 months, it does not make much sense to deter final elevator installation with a hoist that could have been placed in a shaftway that can be patched quickly after an elevator is available to take over the hod hoist's work. Yet every year some superintendent or Project Manager (with the best of intentions) makes this fatal mistake. **You**—don't make that mistake!

26–5   Sensible Temporary Use of Elevators

Under usual conditions, final elevator cabs, which are in most cases fabricated by cab specialists rather than the elevator company, are the last things to be installed before the final adjustments of the elevator system. Once these finished cabs are delivered to the project, they can be installed in a week, and the elevator can be in final adjustment in another week. The big push for elevators is encountered in items 1 through 6 in the work listing in Section 26–3.

Therefore, as soon as the elevator constructors have plumbed their rails, installed and roped the car sling and counterweights, and have made the many electrical connections so that the lifting machine and its controller are connected and adjusted, a temporary *plywood* cab can be erected on the car sling so that the workmen on the building may be moved to upper floors more quickly. As soon as the elevator constructors have completed another elevator to the same state, a plywood cab may be installed on this car sling so that a material hoist can be removed and this temporary elevator used in its stead.

If a building is, say, 20 stories high, there will be elevators that will raise people from the basement to the twelfth floor, and another elevator that will pick up passengers at the first, the twelfth, and each floor above the twelfth. Thus, as soon as this building's structure reaches the fourteenth (low-rise elevator machine room) floor, the low-rise elevators should be well under way so that (as soon as possible) there would be a temporary man-lift cab on which the workmen can ride to the twelfth floor and walk to upper floors. At the same time, elevator constructors should be working in the high-rise shafts (with planked protection above them) so that one high-rise car will be available as soon after its machine-room level is poured.

As wages rise, there are buildings that cannot wait until the first low-rise elevator may be put into temporary operation. In such cases a special type of temporary hoist, much like the "hod hoist" but with special safety features, is installed outside the building's facade to raise the workmen quickly, and thus avoid the wage loss incurred when men have to walk to higher floors. Here, again, this hoist should go outside the building or in a special void left in the slabs, but **never** in a regular shaftway!

26–6   Where Do We Put the Doors?

First, you don't call an elevator doorway a doorway. You call it a "hatch." However, you cannot correctly locate this hatch until the elevator cab's traveling rails are plumbed, because, when the elevator cab reaches a floor, the horizontal distance between the sill of the elevator hatch buck must be precisely located from the sill of the elevator's inner doors. And, as other door bucks may be on the same partition line as the elevator hatches, they cannot be set until the partition line is designated or located by the elevator rails.

Therefore, even if the Construction Superintendent is not too concerned when the elevators *will be running* (which he most surely should be), he **must** be concerned as to how soon the guide rails of the elevators will be installed and plumbed so that the hatchways and all other door bucks for the core may be located and erected.

The installation of elevators has an effect on the entire progress schedule; the watchwords for elevator construction are "as soon as possible!"

26–7   Hydraulic Elevators

219

Motorized
Stairways—
"What's the
Problem"?

As we previously noted, hydraulic oil pressure has now supplanted the older water-pressure system. The older water-piston elevators were rapid, even if they did not have the graduated acceleration that the automatic electric elevators provide. However, the depth of the cylinder excavation and the need to install a piston that would not bend or laterally distort when the car was at its higher floors was costly.

However, oil-hydraulic elevators have an undisputed place in the market for low-rise cars, especially when heavier loads are necessary. The fact that vertical movement is slow because the movement is provided by a small hydraulic pump is not a real problem, because the cars do not travel a great distance. And because the rate of travel is less than in cable-lifted elevators, the control system is less complicated and less prone to malfunction. Oil-hydraulic elevators are slower movers, but they are dependable.

Hydraulic elevators have some of the same problems as the cable-supported elevators. Hydraulic elevators have piston cylinders and guide rails that must be installed and plumbed before hatchways can be installed and before any other door bucks on the same partition line can be located. However, they do not have any counterweights, they do not have direct-current lifting motors that must be powered with AC–DC converters, and their controllers are less complicated. Therefore, because their lifting machinery is fairly standard and there is less of it, the shop drawings are more simple and the lead-time for equipment delivery is less than is required for the cable-lifted electric-motor elevators that are now considered conventional. Their most important requirement is to get the piston cylinder installed before the superstructure is erected!

26–8   Motorized Stairways—"What's the Problem"?

Well, basically, motorized stairways have the same shop drawing and supply problems as do the other vertical transportation systems. Thus no accurate openings in floors can be formed until the shop drawings are finally approved.

But then we have a greater problem in the actual manufacturing. An elevator machine system for a 20-story elevator is pretty much like the system for a 22-story system. Thus, once the elevator manufacturer has a signed contract, he can start making the lifting machinery and its controllers. But motorized stairs have a problem because of the floor-to-floor distances involved. This problem is not one such as a conventional stairway would have, wherein the riser height must be precise to ⅛ in. each so that the number of risers will exactly fit the story height. In a motorized stair the risers are standard (usually 10 in.). The "truss" is the more important part. As the vertical height that a moving stair serves is increased, the length (and strength) of the supporting (slanted) truss must be increased. These truss sections must be designed for every installation, or at least truss sections must be designed for each "bracket" of length. Because of this fact, truss sections for motorized stairs are not shelf items, as are elevator rails. They must wait upon approved shop drawings and fabrication.

There is another hazard in motorized stair systems. Once they are installed and turned over to the owner—no problems. However, if they are finished much *before* they are to be used, their shiny metal surfaces must be protected from the feet of workmen who might walk up them while they are dormant or from materials that may fall upon them. The protective materials used by the elevator constructors are very effective, but often *very inflammable.* A cigarette butt may start a fire that will set the building's completion date back 6 months. Most Project Superintendents would rather have any equipment, including motorized stairs, completed ahead of schedule rather than at the last moment (when protection might not be necessary). However, in the case of motorized stairs, the area must be roped off and placed "off limits" to all workmen.

Chapter 27

Safety: The Lifeblood of the Workmen and the Project

Throughout this text we have interspersed the idea that a good safety program is very important to every project. In this chapter we shall expand on this truth. The title of this chapter tells the whole story. Whereas a good safety program was always helpful in the protection of workmen, a good safety program may now be necessary to save the life of a project.

Insurance rates have always been better for the contractor who had a better safety record. Understandably, broken bones (which lead to lost time) cost the insurance company money, but loss of life costs much more. Each company is rated every 3 years according to the accident experience of the company. If the accident experience is normal, the contractor draws the normal insurance rate for each trade that he hires. If his accident experience is better than normal for the past 3-year period, he is given a certain percentage of *credit* on his insurance rates for the next 3 years. If his accident experience is poor, he will be given a certain percentage of *debit* on his insurance for the next 3 years. Consider a concrete subcontractor with a 10 percent debit bidding the same project as a concrete subcontractor with a 10 percent credit. Even if the contractor with the safer experience were 5 percent higher in his base bid, the 20 percent differential between insurance rates would probably give the bid to the higher bidder who has the lower insurance rate.

One must understand that the overall safety program of the General Contractor has the greatest impact on the safety experience of the subcontractors. Thus, until lately, some large General Contractors who *subcontracted most of the work* were less interested in safety programs than were those who actually completed a greater percentage of the basic work with their own workers. The former type of General Contractor erroneously thought that he had little to do with the 3-year safety record of subcontractors, and, as he was choosing the low bidders, he could not care less. However, this type of thinking is wrong, even if you are only considering profit and not the pain and lost time of an individual worker. First, there is a good possibility that an injured workman will sue a General Contractor for negligence in the case of his accident, and if all construction projects were kept safer, the cost of subcontracts in general would tend to be comparatively lower. Thus most contractors now realize that good safety practice is eventually gainful to all concerned.

When we discuss safety, we mean safety for *all*—safety for the workers on the project and safety for people passing the project. When a project is located near a street where pedestrians could be hurt by a falling brick or other piece of construction material, a construction organization should protect these pedestrians by erecting a bridge which will cover sidewalks that are close to overhead construction. A typical sidewalk bridge is shown in Figure 27–1. The usual location for sidewalk bridges was shown earlier in Figure 5–1.

Figure 27–1. When a construction project is located near sidewalks or pedestrian traffic, sidewalk bridges are required.

27–1 Safety Rules, Safety Laws, and Their Enforcement

Safety rules and safety laws are listed below in the order that we feel they are important. If a construction company ignores items 1 to 3, items 4 and 5 will no doubt force compliance. The sequence that we favor is the following:

1. The laws of common sense.
2. The contractor's own safety rules and safety program.
3. The safety rules of the trade unions on the projects.
4. The safety rules of the State.
5. The safety rules of OSHA.

Primarily, it is important that the reader understand that we are *not* saying that the safety rules of the State and OSHA are not important. What we *are* saying is that if the General Contractor, the subcontractors, and the workers on a project all subscribe to common sense safety precautions, and if **all** the workers on a project combine their individual efforts into a good common sense safety program, the requirements of the State and of OSHA will already be covered. If there is complete cooperation in a project's safety endeavors, it will be a much safer project.

27–2   The Laws of Common Sense

Actually, the basic rules of most safety programs and safety laws are based upon common sense. If there are 200 men working on a project, there should be 200 "Safety Engineers" on the project. If every person on a project kept his eyes open for safety hazards and either corrected the hazard himself or took the problem to the Project Superintendent immediately for correction, the project would be safer even without the regulatory laws. If each worker looked at the project with the feeling that the safety hazard he helped correct would save his own life or the life of a friend, if each worker on a project kept his eyes open for safety hazards, replaced barriers regardless of who moved them in the first place and picked up small pieces of pipe or conduit regardless of the fact that he did not leave them, if each worker told the Project Superintendent of the problems that he could not handle, then we would have a fairly safe project. But there are certain hazards that a worker might see but would not *realize* were hazards. It is here that the contractor's safety rules and safety program take over.

27–3   The Contractor's Own Safety Rules and
Safety Program

Some General Contractors always administered good safety programs from the time they were instituted (only) for the purpose of protecting workmen from pain and lost time. At one time there were two nationally known General Contractors whose safety programs were, comparatively, like night and day. One General Contractor put little money in his project budget for the installation of safety rails and safety precautions, and was happy when its Project Superintendent saved some of this meager budget. The other General Contractor put much more money into the budget for safety. In addition, when monthly cost reports showed that a Project Superintendent was not using a sufficient amount of his safety money, he was called upon to explain to his superiors. Thus, when insurance costs rose, this second General Contractor was far ahead of his competitor insofar as insurance costs were concerned.

In addition to having safety **rules,** a General Contractor or Contract Management organization must have a safety **program,** and someone who checks each project to see that the program is being well administered. The most important part of a safety program is that part which convinces the men on the project that safety is important to them. This is done by holding safety meetings at least twice a month at which the foreman or shop steward of each trade is mandatorily present. At these meetings the Safety Engineer or the Project Superintendent advises all in attendance about safety rules the company wishes observed or where they are not being observed. At these meetings, representatives of the workmen should cite safety hazards they have seen so that the Project Superintendent can have them rectified. Thus, through a two-way dialog, the safety problems of a project are recognized and may be removed.

If, at these meetings, the moderator of the meeting can get the feeling across to the men that safety precautions are the responsibility of *all* the men on a project, he will be making a great advance for safety. For instance, a workman sees that someone has removed a barrier from an elevator shaftway in order to hoist up material. Far too often this workman passes the hazard because *he* didn't remove the barrier, even though a minor effort on his part would replace the barrier. This is short-sighted, and not in accord with the common sense that helps to save lives.

27–4   Safety Rules of Trade Unions

Trade unions were formed to protect their members. Originally, the basic intent of the unions was to procure better wages and more comfortable working conditions. In the last two decades the unions have realized that good working conditions were not possible unless the conditions included *safe* working conditions. Thus a new set of safety rules, union safety rules, have come onto the scene. However, the safety rules of the trade unions and the safety rules of the contractor are basically similar in that they come from the general rules of common sense. Nevertheless, it is important that these unions have their own requirements for a safe project, and that the shop stewards be required to bring safety problems to the attention of the General Contractor at any meetings he holds, or to the Project Superintendent in the event that the General Contractor does not hold meetings.

27–5   Safety Rules of a State

Many States had excellent safety laws (rules) and excellent inspection systems to insure their enforcement. However, *some* States did not have sufficiently strong safety laws and enforcement systems; because of this lack of regulation in some States, the Congress of the United States passed the Williams–Steiger Occupational Safety and Health Act (OSHA) in 1970, which was signed in December 1972. This act took over responsibilities in all States for the safety of workmen in factories, construction projects, and offices, and gave the U.S. Department of Labor the responsibility for a *universal* safety code for all States. We shall discuss this next, but there are good things that the safety departments of several States accomplished that are important to recount so that they may be compared with the systems of OSHA.

A number of our States have had good safety programs and regulations for many years. The rules (or law) were printed and, insofar as *construction-safety* regulations are concerned, were the responsibility of the Project Superintendent or the Project Manager. Under the system of many States, a periodic inspection was made by a trained safety inspector who noted safety-code violations in a written report. Under the system of most States, when a construction company received a list of violations, it had 2 to 3 days in which to remove the violation or receive a fine. In many States the system worked, and good safety results were achieved. However, because there were a number of States which did not have good safety-control systems, OSHA was created.

27–6   The Safety Rules of OSHA

The safety regulations of OSHA need improvement. These safety regulations *have* been revised and will be revised many times. When Congress passed the Williams–Steiger Act and made the U.S. Department of Labor responsible for administering a safety program that would work in all the States and be fair to all employers and all workers, it handed the Department of Labor a Herculean task. Thus there are several cases where, in writing certain safety regulations, OSHA has "reinvented the wheel." We all should realize that **everything** we do (including watching television from a sofa) has certain hazards. In construction there is a certain danger in everything we do. Our problem has been, and always will be, to keep these dangers to a minimum. Mere regulations will not make a safe project. Personal effort is needed. (See Figure 27–2.)

OSHA was created to protect **all** workers in all endeavors. All companies who employ 11 or more workers fall under this act, which is published in the *Federal Register* in part 1910. However, certain industries are covered by special sections. The construction industry is covered by part 1926, "Safety and Health Regulations for Construction," which may be obtained from the Department of Labor or from its offices in principal cities. Every Project Superintendent should have the latest edition.

We have noted that, regardless of regulations, there are hazards in the construction industry and all industries that can only be kept at a minimum. Were we to remove all construction hazards in all construction trades we would have no construction. But a hammer-beaten thumb is one thing, and a head that was bashed in by a falling object (fatal because the worker was not wearing a hard hat) is another! A few hazards, like the beaten thumb,

Figure 27–2. OSHA requires safety rails or ropes around edges of floors prior to installation of facade.

can be controlled by the hammer-holder alone. However, the majority of construction hazards can be controlled by construction supervision and safety engineers. And, for a number of reasons, the requirement of hard hats is one that must be continually policed.

Under the **State** system, when a violation notice was handed to a superintendent with an allowance for several days before a fine was levied, there was good follow-up in some States and not so good in others. OSHA has taken the other extreme. Instead of giving a notice of violation that will result in a fine if compliance is not quickly achieved, the OSHA inspector often levies fines that are final unless appealed within 15 days. And fines for the **same violation** can go to **many people.** For example, if a laborer placed a gasoline safety-can with a defective cover next to the gasoline power-floating machine and then went back to other duties before an OSHA inspector found the illegal condition, he might avoid a fine. However, the cement-finisher foreman, and any other foremen who might be present during the pour (when the OSHA inspection was made), such as the electrical foreman (watching conduit), HVAC and plumbing foremen (installing and watching sleeves and inserts), and the carpenter foreman (checking forms and shores), might **all** receive fines along with the Project Superintendent, who would receive a **heavier** fine. What is the comparative gain over the State system, which gives a warning first?

Surely, if the project is a "slip-shod" operation, a great deal may be gained. Most surely these foremen and their employers, who have been financially hurt, will bring complaints to the Project Superintendent and to any safety meetings, and perhaps the safety atmosphere of the project will improve. However, when fines are handed out to so many people, the costs of subcontracts are increased. In fact, when OSHA first started the "fine" approach, the Building Trades Councils of many major cities feared that subcontractual costs might rise as much as 20 percent. However, there have been some gains to certain people involved in construction, including the Architect and/or his representative. No longer can a Project Superintendent tell an Architect that his warnings about unsafe conditions (which might cost the owner money or, at least, lawsuit inconvenience) are not within the Architect's province. Because OSHA might levy a fine on a **full-time** field representative of the Architect, that representative has every right to protect himself by advising the contractor (in writing when his warnings are neglected) that he has made the warnings to protect all concerned, and that, therefore, the contractor will have to bear the responsibility to repay any fines that may be levied upon him or any other innocent parties to the Project Superintendent's negligence. Thus the Project Superintendent would be well advised to listen to the Architect or anyone else who advises him about unsafe conditions.

Who else gains? The workmen themselves! For example, if the workmen have advised the Project Superintendent that there are unsafe conditions and there have been no improvements in such conditions, they may apply to OSHA and an inspection will be arranged. When the OSHA inspectors come on the project, they will proceed as follows:

1. Conduct an opening conference with the contractor and the architect's representative (if there is a full-time Architect's representative on the project).
2. Make a selection of contractor's employees and subcontractors' employees to accompany the inspectors on a tour of the project.
3. Make a tour inspection of the project.
4. Hold a concluding conference in which the inspectors advise what conditions they find that are in violation of OSHA regulations and what citations the inspectors will make. If fines are levied, those fined with have 15 days to appeal.

There is another advantage. The General Contractor or Project Management organization **must keep** a listing of all accidents that have occurred on the project; this list must show the final outcome of each and every accident. In addition, such records must be kept for 5 years after the conclusion of the project. If accident records are not posted, a fine is mandatory.

It should be noted that under the Williams–Steiger Act any State may take over the responsibilities of the OSHA inspectors if that State can show that safety regulations and inspections under the jurisdiction of the State will be acceptable to the Department of Labor.

27–7   Safety Standards and Their Maintenance

228

Safety: The
Lifeblood
of the
Workmen

Whether the Project Superintendent is acting under his own company's safety standards, those of the State in which the project is being achieved, or those of OSHA, he must have definite standards, and he must maintain these standards. If the project is large, his company may decide to have a Safety Engineer on the project, full time, to make continuous inspections and to chair the biweekly safety meetings. If the project is not large enough to support a full-time Safety Engineer, the company might be well advised to subscribe to the services of a safety-specialist organization, which will periodically inspect the premises and give a report to the Project Superintendent and to his superiors on the inspection and any safety violations found.

Regardless of how the inspections are made, it is almost mandatory that the Project Superintendent arrange for biweekly safety meetings, with foremen or shop stewards of every trade attending. At the first part of the meeting the person who chairs the meeting will remind all in attendance of the basic safety rules of the project, such as the following:

1. Hard hats must be worn at all times.

2. Shaftways may be used to hoist materials from floor to floor if steps are taken to be sure that no one can enter under the hoisted load, and that barriers around the hoistways are replaced immediately after completion.

3. Column-to-column barrier ropes at the periphery of the building shall not be removed by anyone until the skin or facade has been built past that level.

4. No welding or gas burning shall be done unless the welder has a fire watchman with an extinguisher beside him.

5. No open fires shall be allowed at any time.

6. Any dangerous conditions shall be reported to the Project Superintendent's office as soon as they are observed.

The second subject that the biweekly safety meetings should cover are safety hazards that have been found by the Safety Engineer, the Project Manager, or his staff, and safety hazards that have been found by the other attendees. Representatives of the workers will present their complaints here.

Finally, although hazards are mentioned at each meeting, it is not a safe project until the Project Superintendent makes arrangement for their removal and makes arrangements for constant inspections to catch hazards as soon as possible.

27–8   Published Minutes of Safety Meetings

Biweekly safety meetings are very important to insure a safe project. However, if the Project Superintendent has sufficient meetings, makes sure that hazards are removed as soon as possible, and tells all the attendees at the meeting of the importance of removing hazards as soon as possible, he has done but half the job.

In order to **prove** that he has taken every precaution and has advised the representatives of every trade that safety regulations must be adhered to, minutes of each and every meeting must be compiled. These minutes should first list the basic rules, such are listed in Section 27–7, and then record the discussions of the remainder of the meeting. Of equal importance to compiling these minutes is *the publishing* of these minutes. A copy of the minutes of every meeting must go to all attendees, the main offices of all subcontractors, and to the construction company's main office. This is important so that the management of every company involved is kept abreast of the information that is given to their employees.

Of equal or even greater importance, these minutes (and notations or transmittals which show that they were sent to all concerned) are documentary proof that the Project Superintendent or the Safety Engineer *did* in fact do everything in the power of the construction company to show the workers how to be safe and how to keep the project safe, and to show that safety was foremost in the construction company's operation pattern. These minutes may become invaluable if the General Contractor or Project Management organization is sued for alleged negligence in an accident case. Safety precautions are primary. However, unless there are safety meetings and published minutes of these meetings, valuable legal proof of good safety practices is missing. Without that written proof there is considerable chance of a negligence ruling in a lawsuit.

27–9   Repetition of Safety Reminders

Have there been pieces of safety precautions that have been given in one section of this chapter and then repeated (even twice) in other sections? **There are**—and the repetitions are there for a good reason. In fact, throughout this text we have mentioned some of these precautions. This has been done so that the reader will absorb the reasoning that safety is important in **every phase** of construction.

Most surely the primary financial success of a project is of great importance. However, what goal is reached if lawsuits plague the constructor after completion of the project? Thus, whereas monetary profit may be the lifeblood of a construction company, little is gained if damaging suits result from accidents that cripple or kill workmen. Truly, the lifeblood of the workmen and the lifeblood of the project are part and parcel of the project. Keep a safe project!

27–10   Procuring Literature on Safety Legislation

If your project is covered by State safety laws and inspection, you should procure a copy of this law for the field office of the project. This may be obtained by contacting the Department of Labor at the state capitol or at a nearer field office of that department.

If your project is covered by OSHA, you should procure a copy of Chapter 17, Occupational Safety and Health Administration, Department of Labor, part 1926, "Safety and Health Regulations for Construction." This publication lists the forms that must be used to report accidents (especially fatal and serious accidents, which OSHA requires must be reported within 48 hours after a death), the recordkeeping forms that must be posted on the project, and the forms that must be sent to OSHA periodically. This manual may be obtained by writing the Department of Labor in Washington, D.C., but it is usually available at the Department of Labor office in principal cities.

If you desire an in-depth guide to the maintenance of safety in every portion of your construction project, we recommend *Safety in the Construction Industry: OSHA* by Vincent G. Bush (Reston Publishing Company, 1975, Reston, Virginia).

27–11   Particular Standards for Safety

After a Construction Manager, Project Superintendent, or Safety Engineer has read any company safety rules, the State's rules, and OSHA's rules, he should form his *own* set of rules. Whereas these personal rules should cover the minimum situations and standards of official regulations, they should, in addition, be tempered to cover the needs of the particular project. As an example, we all have a special feeling that a 40 mph speed limit on a certain road is "for poor drivers and not for us." The traffic regulation would allow "a good driver like me" 50 mph. However, we all realize that, even though the speed sign says 40 mph, we should drive slower when the roadway is covered with ice. Thus, the roadway covered with ice or the construction situation that is slightly different may require variance from the posted regulations.

Too many Project Superintendents are "construction lawyers" and will not act for safety in a situation unless a particular statute mandates it. They would serve all causes better (including the eventual financial status of their company) if they were to be "construction fathers" who watched over all the men. For example, we have known superintendents that would not place a rail on the edge of a narrow platform 5 ft 6 in. above the adjacent floor because OSHA regulation Subpart M, section 1926.500, states that a railing is necessary for guarding of open-sided floors and runways on a floor or platform that is 6 ft or more above the adjacent floor or ground level. A man can break his leg in a situation where there is but 12 in. in floor difference if particular additional danger aspects are present. Thus adhering to the minimum statutes is not wise. The "construction lawyer" may save a buck for his company for a short time; but, in the long run, the "construction father" will help to keep the company's insurance rate lower and, at the same time, have the personal satisfaction of less "lost-time" accidents and less personal pain on his project.

Official safety regulations are very like the speed regulations on our highways. They cover minimum requirements. The person in charge of a project's safety program must use *personal judgment,* not mere regulations, to insure adequate safety standards for that particular project.

Chapter 28

The Duties of the Architect/Engineer

Chapter 2 was only a formal introduction to the Architect and his Consulting Engineers and an introduction to their part of the work so that we would have a "foundation" for discussing the *Contractor's* work, which is, basically, the purpose of this text. Now, however, we shall discuss the work of the Architect and his team from the standpoint of how it is involved at the time of actual construction.

In Chapter 2 we mention that, in times past, the Architect might have electrical and mechanical designers on his own staff, but that in these days of ever-advancing technologies and specialization, the Architect usually brings a specialist mechanical/electrical consulting firm into the design of a project. Whereas this may have a slight disadvantage in that the mechanical and electrical designers are not in the same office as the architectural designers, there is a balancing advantage in that the Architect may choose the best mechanical/electrical design firm for one project and a different mechanical/electrical design firm for another project which, because there are special requirements, will be served best by this other firm's specialties. And, in addition to having mechanical/electrical consulting firms that are individually better for individually different projects, the Architect has the additional benefit of having his mechanical and electrical design work "spread out" in *several* offices. Thus, tardiness of one mechanical/electrical firm will not be reflected in the design-schedule of another project.

The same situation is reflected in structural designing. The Architect may require basic structural design for one project and on a second project, have an architectural scheme that requires an extremely difficult structural design, where special talents are required. And, once again, the Architect gains because he has several structural designs "farmed-out" in the offices of different structural consultants.

28–1   The Architect's Duties to the Owner

Basically, the Architect is retained by an owner to produce designs and specifications for a building or buildings, and to help supervise the granting of a contract to a General Contractor or a Project Management organization and to represent the owner in such negotiations. When a General Contractor or Project Management organization has been designated, the Architect and his technical staff will check and approve shop drawings, check and approve material samples, and follow the project to be sure that it is completed in accordance with the design-intent of the drawings and specifications and in accordance with the contract between the owner and the contractor. A few years ago the Architect might agree to "supervise" the construction or have his full-time field representative "supervise" the construction. Now, however, owing to legal complications, which have placed unfair legal responsibilities on Architects and their insurance companies, most Architects describe their field duties or those of their representatives as "observing" construction rather than "supervising" construction.

Primarily, the contract for a building or project is drawn and executed between a **contractor** and the **owner.** Although the Architect may have had a great part in choosing the contractor and supervising the contract that was signed, he is not a party to the contract, even though most contracts state that he is the representative of the owner. Thus, whereas the Architect can and does observe the working of the contractor and interpret the "contract documents" (i.e., the plans, specifications, and the contract), he can only *recommend* that the owner take certain enforcing actions (when necessary), or write to the contractor and say that unless the contractor improves certain conditions within a certain number of days (usually three) he will recommend that the **owner** take certain self-protective action. In the old days an Architect or his representative might write that "you are herewith given three days' legal notice to do thus and so and, if you do not, we will stop the project." Sometimes it worked, but it was not legal. If the project actually had stopped, the owner might be legally responsible for delays to which he was not a party.

Regardless, the power of the Architect is great. And, if the Architect uses his power fairly and knowledgeably, he can be sure that the contract documents are adhered to, the construction completed on time, and the owner well served. By the time that construction is started, all the architectural, structural, mechanical, electrical, and special equipment design drawings should have been completed, intercoordinated, and coordinated with a set of specifications that describe designers' requirements for each material or piece of equipment that is to go into the project. Furthermore, the specifications should supplement the information shown on the drawings and make any necessary *written* instructions so that the "intent of the design(s)" is clear. Sometimes this is not always the case, or because certain field conditions are not as expected, certain design changes or at least clarification drawings must be made. The Architect is responsible for these extra designs and/or drawings; in this legal context, the term "Architect" includes all the designers under his direction, including the mechanical/electrical consulting firms and other consulting firms involved in the project and under contract to him.

28–2   The Architect's Duties to the Contractor

234

The Duties
of the
Architect/
Engineer

As soon as the main contract is let and subcontracts are let, many of the contractors will begin submitting shop drawings to the General Contractor or the Project Management organization. After they have been date stamped and recorded in the General Contractor's shop-drawing log book, they should be checked to insure that they are generally correct (i.e., they show the correct materials and correct locations), and then sent to the Architect to be checked. When the Architect receives them, he will also date stamp them, and forward them to his own shop-drawing checkers or will forward them to be checked by his consultants if the shop drawings are other than simply architectural in nature. In some cases shop drawings will have to be checked and/or reviewed by more than one consultant; **always** they should be superficially checked or reviewed by the Architect's checkers *after* they are returned to the Architect and before they are returned to the contractor.

Just as the contractor logs in all shop drawings before they go to the Architect and as they are returned from the Architect, the Architect's staff will keep an accurate log of the date of arrival from the contractor and the date it was sent to be checked. This shop-drawing record should be continuously reviewed to be sure that a consultant or one of the Architect's own staff is not allowing a drawing or set of drawings to be sidetracked or otherwise delayed. This continuous review insures that contractors' shop drawings will be returned to them as soon as possible after they have been submitted. The submission and handling of samples should have the same care and regard. If the Architect *does not* keep shop drawings and samples continually "moving," he will get letters from the General Contractor or Project Manager that he is remiss in his duties to the project and to the contractors. Such letters often become the basis for "extra" claims or claims for delay allowances.

When the Architect has reviewed the shop drawings (or samples), he will place a rubber stamping on the drawing and he will initial the rubber stamping, in addition to noting the date stamped and usually a code number to indicate how this submittal is to be filed. (See Fig. 28–1.) There are four usual notations that will be shown by this rubber stamping:

1. APPROVED: This indicates that the Architect (or his consultant) has checked this drawing[1] and that permission is given to proceed on the basis of this drawing.

2. APPROVED AS NOTED: This indicates that, after checking, the Architect finds that the submission is *generally* correct with exception of minor differences which he has marked onto the shop drawing. Under most specifications, the contractor may proceed on the basis of the corrected shop drawing, but must resubmit the corrected shop drawing for record purposes as soon as possible.

[1] Even though the Architect or his consultant places an approval stamp on a shop drawing, the wording on the stamp usually restricts his approval to "Approval for design-intent only," and leaves the responsibility for actual measurements, adaptation to use, etc., to the contractor involved.

3. APPROVED AS NOTED—RESUBMIT: Under most specifications the contractor MAY NOT proceed until the corrections indicated by the Architect have been made to the shop drawing, and the shop drawing has been resubmitted and approved. Sometimes contractors do proceed before receiving final approval or an "Aproved-as-Noted" stamping, but they do so at their own risk.

4. DISAPPROVED: This marking usually goes upon shop drawings or samples that are totally incorrect, and a notation is placed on the shop drawing (or sample's tag) to explain the reason for disapproval. Obviously, the contractor must make a correct resubmission, as required, before proceeding.

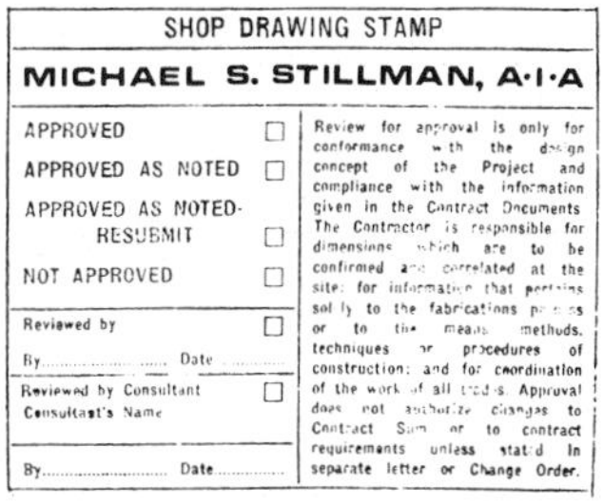

235

Authority
That the
Owner
Passes to the
Architect

Figure 28–1. Architect's shop drawing approval stamp.

28–3 Authority That the Owner Passes to the Architect

The Architect has designed a project for the owner and has specified (or described) the materials that must go into its construction. In many cases the Architect has suggested a number of contractors, who were "invited to bid" the project; usually, he has advised the owner on the type of contract that should be written between the owner and the contractor. And although the Architect is not signatory to the Owner–Contractor agreement, he is always mentioned as the agent or representative of the owner.

Obviously, the owner looks to the Architect and his consultants to evaluate the construction as it proceeds and pass judgment on to the contractor. The Architect must check each shop drawing and each type of material that goes into the construction to be sure that it meets the requirements of the drawings, the specifications, and the design-intent. And, primarily, the Architect and his structural consultant must observe foundation conditions to be sure that the soil (or rock) encountered is equal to that which the designers believed was present. In alteration-situations, the Architect must check to see that existing conditions (when bared) are as anticipated. As the project progresses, the Architect will advise the owner if the contractor's monthly bills (called requisitions) are fair. Obviously, the powers given the Architect are great.

However, inasmuch as there is no contract between the Contractor and the Architect, there are restrictions to what he may order the Contractor to do or to desist from doing. Undoubtedly, if he advises the owner to withhold payment for good and sufficient reasons, the Owner will withhold these payments. And if the Architect advises the Owner that there are good and sufficient reasons for the owner to stop the contract, the owner (after consultation with his lawyers) will stop the contract. But, because the contract is between the Owner and the Contractor, the Architect may not set these actions in operation, but must *recommend* that the owner set these actions in operation. Regardless, even though the Architect does not have legal powers to set certain restrictive actions onto a contractor, most contractors realize that a letter from the Architect stating that he **will** advise the Owner to take restrictive action will usually bring the desired results. Thus the powers of the Architect are great if they are correctly used. If the Architect chooses to be an unfair tyrant rather than one of the people cooperating in the endeavor to give the Owner what is his just due, he will soon be recognized as such and he will not be greatly useful to the owner.

28–4   Duties and Responsibilities of the Architect's Representative

Quite often, when a project is large, an Architect places a full-time representative on the project. Although it might seem that this man was on the project as an "enforcer," the opposite is usually the case. There are many lines on a blueprint and many field conditions that are not anticipated. The full-time representative of the designer has his hands full just trying to **help** the contractors produce the project to the design-intent and under new existing conditions if the project is a large one.

In Chapter 2 we stated that a good Architect was a rare combination of an artist who can design a structure that is both good looking and structurally adequate. By contrast, a good Architect's Representative is a person who has the structural knowledge to know when a portion of the structural design is in danger of being violated during the erection, has the esthetic taste to recognize what is esthetically important to the designer, and by past field experience has a knowledge of what can or cannot be achieved by normal building techniques. In addition, a good Architect's Representative must be 50 percent builder, 50 percent lawyer, and 50 percent problem solver. The reader may say that this adds up to 150 percent. Your author explains that this is so presented to show that the Architect's "Rep" must have many talents, and must be willing and able to put in enough overtime to insure that his duties and his projects are well handled.

Legally, the Architect's Representative should observe the construction of the project to assure himself that it is being constructed in accordance with contract documents and design intent, take necessary action to be sure that the project is being completed according to schedule, receive and take notice of all tests, such as concrete "breaks," structural steel mill tests, and assembly inspection records, interpret all drawings and specifications when so requested, chair all Owner/Architect–Contractor progress meetings, writing the minutes thereof, and keep a legal day-to-day log book. The Archi- tect's Representative's authority and duties are limited as follows:

1. He shall not enter into the contractor's area of responsibility.
2. He shall not advise the contractor how to achieve final results.
3. He shall not suggest or authorize deviations from the contract documents.
4. He shall not personally conduct any tests.
5. He shall not expedite the work for the contractor.
6. He shall not suggest that the Owner occupy the project prior to substantial completion or authorize such occupancy.

In addition, the right or duty of an Architect's Representative to issue certificates of payment or to conduct the required Architect–Contractor or Architect–Owner correspondence is questionable in many duty listings, including those of the American Institute of Architects (AIA). Whether the Architect's Representative should enter into such negotiations and documentation is dependent on the particular representative's ability and legal knowledge and upon the discretion of his employer.

In addition, whether there is a full-time Architect's Representative on the project or a week-to-week visiting representative, this person must observe the project and insure in his judgment that negligence of others shall not result in lawsuits or other financial burdens on the Owner or the Architect. For instance, if unsafe conditions cause a worker to be killed, the worker's family may sue the Owner and the Architect in addition to suing the contractor. And it does not make much sense to observe a project brick by brick (or partition by partition) only to have a ravaging fire ruin it just prior to the Owner's occupancy schedule. Fire insurance is little condolence to an owner who was buying a new, modern structure so that his manufacturing or business costs would be greatly lessened. Not too many years ago a contractor could tell a complaining Architect's Representative that the safety of the workers and the safety from fire, for instance, were the responsibility of the General Contractor (or Construction Management organization), and that the Architect's or Owner's Representative should not interfere. In recent years, however, more strict state safety and OSHA regulations (which we discussed in Chapter 27) mandate that the Owner and Architect complain immediately when they view unsafe conditions, and common sense mandates that they **document** these complaints if resistance is met.

28–5   The Relationship Between the Architect's and the Contractor's Field Representatives

Let's face it. The Architect's Representative and the contractor's representatives will be together for many months. The Project Manager and the Project Superintendent are responsible for the actual construction of a project. The Architect's Representative is responsible to the Architect and Owner to observe that the work is being done correctly and to advise when he feels that it is not. Also, he is charged with the duty of interpreting the plans and specifications when this interpretation is necessary or requested. Legally, the Architect's Representative is on the opposite side of the fence from the contractor, if one considers that there is a "fence." Actually, though, the Architect and the contractor should try to become *one* team, and endeavor to build the project for the owner as well and as quickly as possible. Petty arguments or disputes over authority do not help this primary endeavor. Therefore, a project runs better if there is full *cooperation* between the Architect's and the contractor's representatives.

Nevertheless, all concerned should realize that the Architect's Representative has duties and legal obligations that are different from the contractor's representatives. The Architect's Representative can help a contractor in making a decision, but he may not tell the contractor how to do his work or give any orders to the contractor's foremen.

The full-time field representative of an Architect should have his own private field office or shanty and his own private phone. On some large projects, the A/E firm has a number of persons in its field office. However, regardless of the size of the A/E's field staff, the office should be separate from the contractor's. This helps both the Architect and the contractor to keep their business affairs separate. However, both parties are concerned with the same project and should mutually cooperate whenever possible and within the legal parameters involved.

In summary, the Architect and his representative should (and usually do) try to cooperate with the contractor from the start of a project to its conclusion. However, cooperation is not a "one-way street." Cooperation demands equal effort from both sides. Thus, if a contractor is deviating from contract documents or from general good practice, the Architect **must** take decisive action. In short, a construction project embraces the combined efforts of several teams. If each team does its best at all times so that the owner is well served and quickly, the work of each team will be lessened to the eventual gain of all concerned.

Chapter 29

Changes in Scope of Contracts and Change Orders

Theoretically, if all the assumptions the designer made when he started on the design of a project were correct, and if neither the Owner nor the Architect changed his mind after the contract for the construction of a project was consummated, the project would be completed with the same set of plans and specifications, and the cost would be as noted in the original construction contract. However, this is rarely the case.

29–1   Possible Reasons for Changes

As the actual construction process proceeds, there are changes in the scope of the general contract and/or in the subcontracts. Reasons for changes are many. A few of the normal reasons are as follows:

1. Differences in existing conditions at the site.
2. Changes in the owner's requirements.
3. Changes due to coordination problems.
4. Changes due to the unavailability of certain materials or equipment.
5. Mistakes.
6. Out-of-phase construction.

Most changes in a building program make a change in the cost of the building or project. However, even if a change does not cause a *difference in cost,* permission for the change *must* be documented. In certain cases a subcontractor will be the one who causes the need for a change and he will pay for the change in cost. If the change causes the need for *another* subcontractor to be paid extra money, then his subcontract will be altered by a "Contractor's Change Order" to reduce his original (total) contract by the amount of the *second* subcontractor's cost and a Contractor's Change Order will be made to the second subcontractor to increase his original (total) contract by an amount that will cover his costs for the change caused by another. In such a case the two change orders would be equal in amount and no cost would revert to the owner. In many such cases, however, written permission from the Architect for the change might be required. We shall first discuss the six examples of many reasons that cause changes in contract scope, and then we will discuss how they must be handled.

29–2 Differences in Existing Conditions

Quite often, in the first portion of a construction program, the contractor and architect find that certain conditions on which the design was based are different from those anticipated. For example, if the borings indicated that the structural designer could use spread footings on a soil that would support 10 tons per ft^2, and upon excavation the soil at the elevation shown for bottom of footing will only support 6 tons per ft^2, either the area of the spread footing must be increased so that it will spread the load onto sufficient soil to support the building's load, designated for that footing, or the excavation may be lowered for that footing if there are indications that better support is not too much deeper. Regardless of the answer, a change must be made and, depending on the terms of the contract, more money will be added to someone's contract.

Another example would be when a new building must be connected to an existing building. During the excavation process for the new building, a greater amount of shoring than anticipated for the old building may be required because certain existing foundations are not as shown on the original drawings. Or, as the new building rises, certain structural steel on the older building may need reinforcing. This will cause additional cost and the need for change orders. Usually, costs for changes of this nature revert to the owner.

29–3 Changes in the Owner's Requirements

Let us assume, for an example, that a company is building a larger new building to house manufacturing currently being accomplished in three local smaller plants, so that segments of a finally assembled product may go from section to section by conveyor belt or cart rather than being moved from plant to plant by trucks. In addition to bringing the supervisory offices from three old plants into the offices of the new plant, the owner has left room for future expansion in case he wishes to expand his plant or to handle the paperwork of an out-of-state plant from this new home office.

During the construction phase of the new plant, the owner may find that certain new machinery that he had intended to use has been changed in design or size, and that the foundations or bases for the electrical service for this new machinery must be changed. This will require redesign, and "bulletins" from the Architect and General Contractor will issue change orders to subcontractors. During the construction phase, the owner may find that he wants to bring all the office workers from his out-of-state plant to the home office (or replace them with local people) and expand the new plant's office areas. Here again is a change often encountered. The changes in office expansion in the new plant will require new design and bulletins from the Architect and change orders from the General Contractor or Construction management organization. These would be owner-paid changes.

29–4   Changes Due to Coordination Problems

As we noted in Chapter 25, most contract documents require that the mechanical/electrical subcontractors coordinate the placement of their trade's equipment and lines with the space requirements of the structure and location requirements of the other trades. However, basically, you cannot put more than ten pounds of kumquats in a ten-pound can! Sometimes the Architect has not given enough space in certain areas. In turn, the mechanical/electrical consultant has not tried to persuade the Architect that more space is required, because he does not wish to displease the Architect or he feels that "there is some way" to fit in all the machinery, ducts, and piping into the space if a Spartan effort is made by the contractors. If this is not possible, changes will be required to increase the size of this area. In such a case there will be change orders that will increase the cost to the owner.

29–5   Changes Due to Unavailability of Materials
or Equipment

Quite often, after a project has been designed and specified, market conditions or manufacturers' product changes make certain materials or a piece of equipment unavailable. The equipment of another manufacturer or a new model of the specified manufacturer may cost more or weigh more. If there is extra weight, the structure may have to be strengthened so that it can support the heavier equipment. If the new equipment causes costs that are higher than the originally specified equipment, someone will have to pick up the extra cost. The determination of who pays for such extra costs usually depends on whether the change was made before or after the bidding period. If the change was made before a subcontractor bid the project, he should have included the change and should not expect to be paid extra. However, if the unavailability of a product presents itself *after* he has been awarded a subcontract, he can usually depend on extra payment if a change causes him extra costs.

Everybody makes some mistakes! If we didn't there would be no erasers on pencils. Of course, when an owner is paying an Architect and consultants to design a structure, he depends on them to design without mistakes or omissions. Maybe these designers do not make mistakes, but sometimes they forget to investigate certain possibilities that might later cause a change in construction. In either case, there will be extra monies that must be paid by someone. Obviously, it is not the contractor's problem, and he will wish to be paid for the extra cost. If the Architect and his consultants have generally done a good job in the total design, the owner will usually agree to absorb this cost. However, if the owner feels that there has been gross negligence on the part of his designers, he may wish to have them contribute to or repay him for the entire extra cost of this change.

Sometimes a contractor or a subcontractor makes a mistake in installation preparation or omits to prepare for some portion of his installation. For example, if the electrical subcontractor stubs his conduit down incorrectly so that it misses a partition, and the owner does not wish an exposed conduit line, he may have to pay to have a masonry pilaster built around his conduit. If an HVAC subcontractor miscalculates and runs a steam line in a location where a hole must be cut through a girder and reinforcing plates welded to that girder, **he** will have to pay for that steel work.

29–7 Changes Caused by Out-of-Phase Construction

It is generally good practice to have a project completely and finally designed and specified from foundations to roof before the project goes to bid. This has always been true when there was little national inflation and cost-index graphs were fairly level. Even when cost factors in building construction rose almost one percent per month, it was usually good practice to design first and contract afterward. However, as soon as inflation causes building costs to rise two or three percent per month, out-of-phase construction makes more sense and is practiced more often.

Under the out-of-phase system, contracts for a building are let as soon as there is enough design for the first phases of a project to go into purchase and construction. Of course, this will surely bring "extras" for items that did not show on the original plans or are changed after a contract is signed. However, in inflationary times when a building is delayed six months until the final plans and specifications are completely finished, the general local inflation may add 12 to 18 percent to a project's cost if one figures a price rise of two to three percent per month for six months.

On the other hand, if contracts are let as soon as a contractor or subcontractor has enough information to order materials and machinery, the contract bids will be lower. Thus it would be better for the owner to pay for four to eight percent of the originally figured cost in extras and to be able to take beneficial occupancy a few months earlier than to wait until all designs were finalized and pay from 12 to 18 percent extra. This type of construction has become more prevalent as national inflation has increased, and is one of the factors that encouraged the development of Construction Management organizations. These organizations work very much like a General Contractor on a "Cost-Plus" contract would work. However, as Construction Managers, they usually take fewer risks even though they have less profit incentive.

29–8   Change Orders Issued by the Architect

If the Architect issues a revised plan with changes, he will usually send it to the General Contractor along with a bulletin on which he lists the drawing numbers and lists the changes involved. This bulletin will request a change in price (either add or deduct), and will note whether the contractor should proceed immediately or wait until the new price is approved. After the contractor has submitted the amount of this change along with back-up material, and perhaps after additional Architect–Contractor negotiations, the Architect will write a change order, sign it, and send it to the owner with a letter of recommendation. When the owner signs the change order, he removes and files his copy and sends all other copies to the Architect for distribution to the Architect's files and to the contractor.

As the project continues, there may be changes that the Architect did not contemplate. He may be notified by the contractor that there is cause for the contractor to receive extra monies due to conditions he did not consider when he made his bid (or when a subcontractor made his bid), and that he wishes a change order from the owner. If he has a basis for this claim, and after he and the Architect have agreed on a fair amount, the Architect will write a change order and forward it to the owner along with an explanation and recommendation. When signed, this change order is processed as above described.

29–9   Change Orders Issued by the General
Contractor or Construction Manager

There are two kinds of change orders issued by a General Contractor or Project Management organization: (1) change orders written for subcontractors after the General Contractor has received a change order from the Architect, and (2) change orders that subcontractors receive when the scope of their contract is changed by a decision of the General Contractor.

In the first instance, where the General Contractor has received a change order from the Architect, the work shown on the change order must be distributed to the actual subcontractors who will perform the work. For instance, if the Architect's change order covers the building of an additional room, there will be additional partitions and paint, additional or changed lighting, and changes in the acoustic ceiling, if not more materials and subcontractors involved. The Architect's change order was made on the basis of costs for these subcontract changes as compiled by the General Contractor, which, in addition, included the General Contractor's fee for the additional work. Thus, when the General Contractor has a change order from the architect, he must give change orders to each of the subcontractors who will do each part of the additional work. If the Architect's change order was a "deduct" change order, there would still be a need for adjustment change orders to subcontractors involved.

The second kind of General Contractor–Subcontractor change order occurs when a subcontractor has to do someone else's work or make repairs after someone else has damaged his work. For example, if a bricklayer stood on a plumber's piping while he was finishing the top of a masonry partition and the piping needed repairs, the General Contractor would authorize repairs by the plumbing subcontractor and would "backcharge" the masonry subcontractor for the cost involved. Whereas the charges involved might be held in abeyance for some period, they would eventually show up in a change order to the plumber, which would (for this item) add to his contract price, and a change order to the masonry subcontractor, which would (for this item) deduct from his contract price. We note that this item might be held in abeyance because most General Contractors prefer to hold backcharge items until there are a number that can be included (both add and deduct perhaps) on one subcontractor change order for the subcontractors involved.

The first kind of change orders, those which distribute changes made on an Architect's change orders, will have a result on the total sum of the general contract. The second kind of change orders, those which make charges against one subcontract but make equal additions to other subcontracts, make no net change in the total sum of the general contract, but merely redistribute the funds.

29–10   Emergency Field Orders

Under all the rules of good practice, changes should not be made until they are authorized. However, there are times when work must be done immediately, and the cost of the work figured and agreed upon later. An example of such a situation would be when someone realized that new equipment would place a heavier loading on a portion of a slab that was to be poured the **next day.** In this case the Architect would confer with the structural consultant to determine what additional reinforcing steel was required, and might even have time to get a location sketch for the added reinforcing bars. However, there might not be time for the contractors to estimate the probable time and material involved in the extra work so that the architect could make up a formal change order and have it processed. In this case the Architect would give the General Contractor an "**Emergency Field Order**" to have the work done and to present the charges (along with back-up) at a later date when a formal change order would be written.

In the same way, the General Contractor may have to give emergency orders to his subcontractors to perform extra work and determine the cost later. The work involved may result from an Architect's emergency order, or it may result from work that the General Contractor cannot do with regular forces at this time.

Most Architects and General Contractors do not wish to give emergency orders. They would prefer to handle changes in the normal manner. However, there are times when the project cannot wait for formalities. If the men involved know their business, they should be qualified to give a verbal order and follow it up immediately with a written record so that there will be no disputes at a later date. To state the matter differently, requests for changes should be kept to a minimum and orders should come from the main office. However, when immediate charges are necessary, *correct* emergency instructions must be given immediately.

One warning, however. Verbal orders (without written confirmation) should not be allowed. As soon as a Project Superintendent hears an extra request from an Architect or an Architect's representative, he should ask, "who is going to pay the extra cost?" If he does not, his office will have him "on the carpet" shortly. Project Superintendents will do themselves and the Architects a service if they make it a practice to be sure everyone involved is aware of "who pays" every time a verbal extra request comes up. In the long run, this habit will save a lot of misunderstandings and might even save someone's job.

Then, too, there are changes that may not cause extra costs and some that may even save money. Regardless, if there is a change, it should be covered in writing so that all concerned have a record.

Chapter 30

Records and Payments on the Project

From the time that actual construction of a project is started to the time that the owner moves into the project and "takes beneficial occupancy," many records must be kept. Even after the owner takes beneficial occupancy, there will be records of remedial work required by the Architect and, of course, payment records. On the modern project, which may have typists and secretaries right in the field office, many records will be in the form of project correspondence and the change orders discussed in Chapter 29. However, there are certain things that every Project Superintendent must see to *regardless* of the size and scope of the project.

30–1   The Daily Log

Every construction organization has a printed form for the daily log of the project, which notes the important day-to-day facts (see Fig. 30–1). Each construction company has a slightly different printed form because each company has a desire for special information for its records. Basically, however, there is a listing of the weather conditions (including temperatures) for each day. This is a very important item, because extensions in time or payment for extra cost may depend on proof that greater-than-usual rainfall or extreme weather conditions were experienced during the contract period that delayed the project. In addition, all daily log forms have columns to list the number of men for each trade (broken down into the number of foremen and the number of journeymen) and, in the space to the right of each manpower listing, a description of what work these different trades accomplished each day.

There will be a space at the bottom of the form for the Project Superintendent to record any unusual circumstances or special orders the company received that day. In the event that an "unusual circumstance" for one day is a notation of an accident on the project, the superintendent would be wise to confer with his company's main office (in case his superiors feel that a lawyer's advice is needed) before the notation of the accident is finally

Stillman Construction Company

DAILY LOG

BUILDING __Johnson & Company__
LOCATION __Pikestown, N.J.__
CONTRACT NO. __1125__

WEATHER __Clear, cold__ DATE __December 14,__ 19 __76__
TEMPERATURE 8 A.M. __20°__ 12 M. __39°__ 4:30 P.M. __33°__

CLASSIFICATION OF TRADES	FOREMEN EMPLOYED	MECHANICS EMPLOYED	DISTRIBUTION OF TOTAL MEN EMPLOYED TO LOCATION OF AND WORK PERFORMED
S.C. COMPANY Staff	12		1 Project Mgr., 1 Sup't., 3 Ass't. Sup'ts., 1 Field Engineer, 1 Ass't Fld. Eng'r., 1 Plan Clerk, 2 Acc'ts., 2 Secretarys.
" Carpenters	1	4	Installing protection around openings, replacing guard rails at elevator shafts, repairing fences at Main Street entrance.
" Laborers	1	10	Tending carpenters, Placing tarpaulins for protection of next concrete pour. General cleanup operations.
J.C. JONES EXCAVATING CO.	1	8	1 Bulldozer, 1 shoveldozer, 1 Backhoe, 5 Trucks. Cleanup, Bldg. #1, starting excavation for Bldg. #2.
ANDERSON CONCRETE COMPANY Super. Carpenters Laborers Ironworkers Cement Fin.	1 1 1 1 1	12 20 2 4	Building forms for 4th floor, Building #1-(Eastern third) Pouring Concrete for 3rd floor Building #1 (Western third) Removing scrap lumber. Stripping forms. Tending temporary heat.
WILLIAMS PLUMBING COMPANY Super Plumbers Laborers	1 1 1	3 5	Placing sleeves and inserts, 4th floor and tending concrete pour. Excavating for underground drain lines, first floor, and installing these lines. (Note: S.C. Company's representative and Architect's representative observed pressure tests in the M-3 area, 1st floor)
FRANKLIN ELECTRICAL ENG'RS. Super Electricians	1 1	8	Installing conduit on forms for 4th floor (East), Bldg. #1, Tending concrete pour 3rd Floor (West), Bldg. #1. Tending temporary light and power.
WARNER MECHANICAL CONTRACTORS Super Fitters Sheet Metal	1 1 1	4 2	Placing sleeves and inserts, 4th floor and tending concrete pour. Starting secondary water risers and run-outs at 2nd floor. Installing hangers for ductwork at 1st floor. Recing sheetmetal ductwork.
J.F. JOHNSON COMPANY Super Ironworkers	1 1	4	Installing inserts for facade of building #1. Receiving material.
ERB WATERPROOFING COMPANY Cement Fin. Laborers	1 1	1 4	Installing metalic waterporoofing in elevator pits, Bldg. #1.
A-1 SEWER SERVICE (Subcontractor of Williams Plumbing)	1	8	Excavating for and installing new 16" storm sewer line in Main Street.
VIACON FIRE-PROOFING CO.	1	4	Receiving material, setting up machinery for spray-on fire-proofing.
EDWARDS PROTECTION COMPANY	1	6	Completing sidewalk-protection bridges around Building #2.
OWENS ELEVATOR COMPANY	1	4	Setting up field shanties and receiving material.
GUARDIAN WATCHMAN SERVICE		2	Watchmen for 4-12 and 12-8 AM
TOTAL	36	115	

GENERAL REMARKS
1. Messrs J. Franklin and F. Lieb of Johnson & Co. visited project this day.
2. Mr. Fred McNamara, representative of the Architect made his weekly visit this day.
3. Mr. Brian Jones, representative of the Structural Engineer, inspected sub-soil for footings of Building #2 and approved same.

Send Original Copy to Office

John Ashley _____ Supt.

NOTE: Record visitors, also any unusual conditions, such as strikes, etc.

Figure 30—1. Contractor's daily log.

worded and typed onto the daily log. Once this daily log is written, signed, and copies sent to several offices, it becomes a **legal instrument.** A daily log or a series of daily logs may be subpoenaed at a later date for use in a lawsuit. Thus the wording of a daily log should be precise, because it is the cornerstone of proof when a claim is made by or against the contractor. A copy of each day's log should go to the main office daily. Unfinished logs should not be allowed to accumulate. Because a log notes items that may happen right up to quitting time, it sometimes waits for final typing until the next morning. Regardless, it should be in the mail the day after its date.

30–2   The Superintendent's Daily Diary

Another important record is the superintendent's daily diary. Some superintendents keep a diary for themselves from January 1 to December 31st regardless of what project they are assigned to. Thus, over the years, there may be records of more than one project in some diaries. A better system is to secure a new diary volume for each new project. Thus, if a new project starts on May 1, there will be blank pages from January 1 to April 30. The notes that the superintendent places in his daily diary are items that will supplement information that is typed onto the log, will record special orders he may have received that day from a superior (which would not be in the log), or will be special, personal notes that will help him reconstruct a situation at a later date if the need arises. The Project Superintendent should endeavor to place some notation in the diary for every project day, even if there is nothing of great importance to insert so that, at least, the diary will be another continuous record for him to refer to. If the daily log form does not have a requirement for listing visitors and inspectors that were on the project, this information should go into the daily diary. This diary is a *personal* record and, therefore, should not be accessible to everyone; it should be locked in the superintendent's desk or in his personal file case.

30–3   Monthly Progress Payments

Most specifications require that within a certain period after a contractor receives his signed contract (usually 15 days) he provides the Architect with a trade-by-trade breakdown of the total bid cost of the project. This breakdown does not have to reveal any confidential information such as the precise amount of a subcontract. In many cases, General Contractors show larger than actual amounts for primary items such as excavation, foundations, and structural steel, and less than actual amounts for later accomplished items such as painting and site work. The contractor's primary purpose for this front-end loading of the breakdown is to make a slightly greater percentage of the total budget available to the contractor earlier in the project; if the breakdown amounts are not distorted beyond reason, the

Architect and Owner will accept this breakdown. Once the breakdown has been submitted and accepted by the Architect and Owner, there is a firm basis for payments to the contractors.

Thereafter, once each month, the representative of the contractor will sit down with the representative of the Architect and review the percentage of completion that the contractor feels is accomplished in each trade (or subdivision) up to that point. Once these percentages are agreed upon, the contractor's office will prepare a typed monthly bill (called a **requisition**) listing the dollar amounts for each subdivision and the total amount of all subdivisions. This will be the amount in dollars of the total contract that has been accomplished.[1]

If the basis of the general contract is Cost Plus Fixed-Fee, the breakdown will list the actual amounts of each subcontract and, once each month, the representative of the contractor will review each subcontractor's requisition with the Architect. Thereafter, the General Contractor's requisition will list all these amounts (less a retainage on each), show the total of the amounts, and add a percentage of this total as the General Contractor's fee.

30–4   Retainage on Requisitions

Because an owner could never have a project completed by another contractor if he had paid the original contractor for 100 percent of the work that was accomplished by each requisition time, the contract usually stipulates that "until substantial completion of the work" a certain percentage (usually 10 percent) will be retained from the bottom line of the monthly requisitions. This has the effect of retaining the same percentage from each subcontractor and from the General Contractor's "general conditions" (the General Contractor's overhead and profit) portion of the requisition. In the case of the Cost-Plus general contract, there will be at least 10 percent retained from each subcontractor's monthly requisition, which will take away at least 10 percent from the General Contractor's fee.

For a complete explanation of the reasoning on retainers, see Chapter 3. Over the years, it has been general contractual practice to retain at least 10 percent until "substantial completion" of the project (usually 60 to 70 percent) and then reduce the retainage to 5 percent.

30–5   Bonding of Contractors

On a number of projects (especially municipal, State, and Federal) the documents require that the General Contractor and/or each subcontractor be **bonded.** This means that the contractor(s) must purchase bonding insur-

---

[1] In some cases, subcontractors will be required to submit notarized statements that they have paid (all but listed) bills before they can be paid on the next requisition. Similarly, General Contractors are required to state that they have paid all subcontractors.

ance from a company that specializes in such bonding. At such time that a contractor finds he is unable to complete his contract, the bonding company will step in and place another contractor on the project to finish that work. One might feel that, if a contract is bonded, there is less responsibility placed upon the Architect or whoever decides what percentage or progress payment is due that contractor. Actually, the opposite is the case.

For example, if the Architect has agreed that a contractor has completed 60 percent of his work when, actually, he has completed only 50 percent, there may be problems with the bonding company if the contractor goes into bankruptcy. The bonding company may take the position that the owner paid the contractor more than was actually due and, by so doing, took responsibility for the extra 10 percent. If the bonding company were to take this position, it would pay for only 40 percent of the total contract. Thus, even when a contract is bonded, the Architect will take great care in approving the percentage due each contractor.

30–6   Compilation of Change Orders and Their Payment

As the construction of a project progresses, there are bound to be additional change orders that will require payment. Therefore, as these change orders are approved, they will appear on the monthly requisition. The amount due on each change order will be billed by percentage completed just as the subdivisions are billed. The total amount due on all change orders will be added to the total amount due on the base contract, and the regular retainage percentage will be deducted from that sum.

30–7   Final Payments and "Release of Liens"

When an owner pays a General Contractor each month, he has every right to believe that the General Contractor is paying the amounts requisitioned by the subcontractors. But, because this is not always the case, the Architect may require that the General Contractor submit a signed statement with each monthly requisition, which states that the General Contractor has paid amounts previously billed by his suppliers and subcontractors, and that he intends to pay his suppliers and subcontractors up to date as soon as he receives payment on the current requisition. However, if a subcontractor or a materials supplier has not received payments, he may place a **lien** on the project.

The amount of time that the law allows contractors and suppliers to place liens on completed buildings varies in different States. However, these contractors and suppliers are always allowed to file liens (when money is due) for several months after a building is completed. Thus most contracts between owner and builder stipulate that the General Contractor (or Project

Management organization) must secure a **release of lien** from each subcontractor before the Architect will certify that the owner may pay 100 percent for that particular subdivision. Thus, before the Architect will certify that the owner should pay 100 percent of the total contract, he will want to have lien releases from all parties who have contracts with the General Contractor or Project Management organization.

Of course, before **final** payment can be made, the project must be finished to the satisfaction of the Architect and the Owner. We shall discuss these final matters of the contract in the final chapter.

Chapter 31

Labor Relations

No project can be erected without satisfactory labor relations. To cover every phase that must be considered in labor relations is not possible here. But there are basic rules for handling labor problems and a construction job. These are the subject of this chapter. The reader will need more information and will have to be conversant with many other documents if he is promoted to a position where he is in charge of the labor problems for his company. However, for the Project Manager and the Project Superintendent, this chapter's outline should suffice.

Labor agreements between construction companies and the workmen they hire are made between representatives of "Management" and "Labor." Agreements differ from area to area. Thus, if a large construction company has a contract to construct a building in an area outside of its own normal practice, it will have to abide by the union–management agreements of **that area.** However, if the construction company has offices in certain areas, it will be a member of the local section of the Associated General Contractors, a "Building Trades Council," or a "Building Trades Employers' Organization" which meets with the management of the different unions, agrees upon the jurisdiction of each trade union, negotiates with the unions, and signs an agreement (in the name of all member contractors) with each trade union. After these agreements (or contracts) have been signed by both the officials of the contractors and the elected officials of the union, the agreements are printed by each union in a small pocket-sized book so that each union member may have a copy. More importantly, the unions will send copies of the book to all contractors so that the contractors' labor-relations people will have material with which to make their own Project Managers and Project Superintendents aware of each agreement.

The agreements made by each "Local" union cover three points: (1) hourly wages for the contract period, (2) working conditions, and (3) a general agreement on the jurisdiction that a certain Local has for the right to do certain work. For example, the organization for the General Contractors and subcontractors in a certain area may agree with the carpenters Local that the erection of corrugated asbestos-board siding is to be done by carpenters. In another area, the union agreement might give this work to the sheet-metal workers.

Thus, as we have previously mentioned, *certain* work goes to different trades in different areas. However, basically, subdivisions of work go to the same trade union throughout the entire country. Each local union is a member of an international union, and the generally agreed on jurisdictions are printed in the international books.

31–1   Negotiations with Union Delegates

Many large construction companies are so adamant in their company regulations that they will not allow a Project Superintendent or even a Project Manager to deal with unions or even to have copies of the local agreements. The reasoning of such a regulation is substantially sound.

The original agreements were made by the construction company's management with a union or by an employer's association with a union. In some cases the jurisdictional portion of the contract was not arrived upon easily. Therefore, a construction company is reluctant to allow a young Project Superintendent to settle a particular dispute with a union delegate because the new project agreement may set a precedent for *future* union contracts that would not be palatable to the particular contractor or contractors association. However, we do not feel it is sound policy for a construction company to forbid its project organization to have the union contract books on the project. The Project Manager and Project Superintendent should be *aware* of the jurisdictional agreements, even though they are not empowered to negotiate with union stewards and union delegates. If a construction company's field representative is fully conversant with the local trade agreements, he will not be trapped into creating a project dispute by ordering a worker to do something that is, contractually, the work of another union. A jurisdictional dispute on a project can create great damage to the schedule.

For example, in many areas the Dockbuilders' union whose members drive the piles, make the exact pile cuts at the top of the piles, and form the pile caps, have also claimed and have been awarded all concrete forms for a height of 6 or 8 ft above the top of pile caps. Some of the grade beams or peripheral concrete girders that sit upon these pile caps require very fancy carpentry. It would be normal for a conscientious Project Superintendent to put "finish carpenters" onto these forms. There might be very few dockbuilders in a local union who were able to do this work. Most dockbuilders are used to working with heavy timbers, not fancy forms. If the Project Superintendent was aware of the jurisdictional agreements, he might be able to have his authorized company labor representative make a "trade-off" agreement for the project wherein the dockbuilders' and the carpenters' unions (along with the contractor) mutually agreed to allow the dockbuilders certain additional work in trade for the carpenters doing certain of the grade-beam forming. Regardless of the authority of the construction company's men on the project, these men should **know of** the jurisdictional dividing lines so that time-consuming disputes can be avoided. They should have the agreement books.

Furthermore, there are always a few trade stewards who will claim anything. In such cases a superintendent who has all the agreement books and has read them can say, "I don't believe it is in your book but if you can show it to me I will discuss it with my company." Also, if there is a dispute between two trades on the project, he can pull out the "International" books to ascertain if the dispute has ever been settled by the unions' international. Most project stewards do not have these books. In short, if a construction company's site representative has all the agreement books he is in a position to stop a dispute **before** it makes any problem. A construction company has good reason to make sure that its site representative does not make any agreements that would be injurious to the company, but it should not refuse him the knowledge contained in the agreement books.

In cases when a construction company is awarded a contract in a locality where it was not signatory to the agreements with the locals, it is usually good company practice to abide by all agreements made by the local contractors' associations. In such a case it is almost mandatory that the site representatives have copies of all locality agreements. However, because this company was granted a contract to the loss of local competition, it cannot generally count on any local support when labor problems arise. In such cases the main-office labor relations man may have to come to the project and assist in the settlement of the dispute.

31–2   The Amount of Wages Paid on a Project

The rate paid to a construction company's site-management personnel is dependent upon company policy; if management personnel are required to move from one area of the country to another, the amount of extra remuneration is dependent upon general company policy and upon individual negotiations with each individual.

Although a construction company may make special allowances to certain foremen that it sends to a project, it must pay the going rate of the locality to the journeymen it hires from the unions in the locality. It can pay no more and it can pay no less. If the local unions cannot supply enough workers, a construction company may hire union workmen from another locality and may *then* pay certain travel allowances to these men so that they will come onto the project. However, such tactics should not be employed until they have been fully discussed with local delegates.

31–3   The Open-Shop System

The employment laws of many States now mandate that a workman must be hired regardless of whether he belongs to a union or not. However, this system is hard to enforce. A contractor may decide to hire good workmen regardless of whether they belong to a union or not and to pay each man the rate that the union agreement requires. This will fulfill the State statutes, but will it provide him with enough workers? The contractor may be able to hire good nonunion workers for certain aspects or trades, but he will not be able to complete his project if union men will not join the project. Thus, whereas a contractor who specializes in small projects may be able to complete these projects on a local basis with nonunion men on the payroll, the contractor who specializes in larger projects and/or projects in many different States will find it hard to complete a project if there are nonunion men on the project. We do not necessarily feel that this is right, but we would be less than honest if we did not state that it was a fact of life.

31–4   Trade-Union Affiliation

The American Federation of Labor (AFL), which traced its genealogy back to Samuel Gompers and the Knights of Labor, held sway in this country from 1886, and covered the industrial labor unions and, as they were formed, some construction labor unions. The AFL covered most trade unions until, in the first part of the depression of the 1930s, when everyone was trying to find new ways to make a dollar, a number of splinter-group unions were formed. In 1936, when there were power struggles within the AFL, several industrial unions were expelled. Shortly thereafter, they joined with other industrial unions to form the Congress of Industrial Organizations (CIO).

When there was a separation between the unions under the AFL and those under the CIO, most of the building trade unions were AFL. For many years there was a problem with certain equipment, especially boilers and electrical equipment, that was manufactured in factories by CIO workers when it was destined for construction projects manned by AFL workers. For instance, there have been cases where AFL construction men insisted on disassembling and reassembling a CIO-manufactured "packaged" piece of equipment (i.e., a piece of equipment that was fully independent and needed only connection to mechanical/electrical lines) before it could be used on a project. This was a thorny problem until 1955 when the AFL and CIO became one organization, the AFL–CIO.

However, we still have problems with certain "umbrella" union groups in certain parts of the country. For example, the United Mine Workers, which was organized in the latter part of the nineteenth century to bargain for coal miners and iron miners, took over all trade unions in the mine districts. That is, the United Mine Workers carried the machine operators who ran the mining machinery, the truck drivers who drove the mine company trucks, the electricians who did the wiring in the mines and mine buildings, and the construction men who built structures in mine towns, in addition to the actual miners. This was a natural occurrence. In certain areas of southern Ohio, western Pennsylvania, and northern West Virginia, much of the construction of buildings and roads is done by workers belonging to the UMW.

Thus, a Construction Manager who is buying subcontracts must be careful
to be sure what union a company's men belong to before he invites that firm
to bid. True, federal law says that unions of two different parent affiliations
must be allowed to work on the same project, and if AFL–CIO unions
complain when a UMW union enters upon a project, the federal government will obtain injunctions forcing union delegates to instruct their men to
return to work if they have left the project. The delegates will then so
instruct their men. However, they have no control if the men call in to say
they are "sick." Thus, whereas federal laws *can* make different unions work
together, just as federal laws *can* make union men work alongside nonunion
men, it doesn't always happen that way.

The Architect does not and cannot word the construction documents to
force a construction organization to use nonunion or union help or to abstain
from using workers of a certain union. What he can do and often does do is
to word the specifications to make the construction organization use workers
"who are qualified for the work and *compatible* to each other so that the
project will not be delayed." This places the responsibility back on the
construction organization and Project Manager. He must use great caution
and foresight in his buy-out of subcontracts.

31–5   Jurisdictional Disputes

In Section 31–1 we discussed why some construction companies wish to
have disputes arbitrated by home office men rather than field personnel, and
showed how minor disputes could often be settled immediately by checking
a union's agreement book or the international book.

But there are certain disputes that we could almost label long-lead disputes. As we have mentioned in other parts of this text, there are certain
materials or pieces of equipment that the Project Manager should intuitively
know may be claimed by two or more trades. For instance,

1. Who places a portland-cement base paint on a concrete spandrel, the
   painters or the cement finishers?
2. Who sets fiber-glass wall units into Architectural Metal frames, the
   glaziers, the carpenters, or the ornamental iron workers?
3. When a concrete floor has been poorly troweled when originally
   placed, who patches the floor (when very deeply embedded)? The
   carpenters, who claim that they are preparing the concrete for resilient flooring (which is their responsibility), or the cement finishers,
   who claim that they are "repairing" the concrete floor so that the
   carpenters can "prepare" the repaired floor surface for resilient tile?

These are but three of thousands of long-lead jurisdictional problems that
have come up in the last few years and have been solved. However, before
the jurisdiction was ruled, some projects lay still for months. Thus the

Project Manager and the Project Superintendent should go over the plans carefully at the very beginning of the project so that some of these very obvious problem areas can have consultation and solution before the problem installation is required.

31–6   Labor Relations: Recapitulation

A construction project is merely a large crowd of workers doing their own thing. For the project to be a happy project, the workers must be happy with their lot, working conditions, and safety conditions. No project will be successful if the men are not happy, and most projects will be successful (from a labor point of view) if the men and their union delegates feel that they are getting a fair shift.

This is not to say that every union delegate should be granted his every whim or, in fact, *any* whim. It must be in his book! However, the union delegate and the union steward he places upon the project must be assured that their men will work within established union agreements, that no one will steal their work, and that their men will work in safety. In the latter requirement (safety), a good Project Superintendent makes sure that the delegate and the steward are shown that they too must help to keep their men safe.

Finally, the men themselves, regardless of their trade, must be kept comfortable and safe. For example, recently we observed a project where plumbers, electricians, and steamfitters were installing their material and bricklayers were laying up concrete block partitions in areas where rainwater was 1 in. deep on the floor, rainwater from upper floors was dripping on everyone, and temporary heaters were so scarce that the ambient temperature was barely 45° F. A temporary roof six stories above and three times as many heaters would have allowed every man to work in comfort, which would produce better workmanship and nobody would lose time because he caught pneumonia!

Labor relations, like one's own relations, must be treated carefully.

Chapter 32

Finishing the Project

When a builder signs a contract to erect a structure or group of structures, he agrees to build the structure in full compliance with the design drawings and in full compliance with the specifications that are a part of his contract documents. The general conditions portion of these specifications requires that, in addition to erecting the structure in accordance with design-intent and with materials described in the specifications, he obtains a building permit from the town or municipality and that he completes his work in full compliance with local and State codes. Thus, when a builder feels that he has completed his project, his work will be judged by the following:

1. The Architect/Engineer, who will inspect the final project to be sure that it complies with his design.
2. The Building Department of the town or municipality whose inspectors will judge whether the structure has been built in compliance with local and governing State codes and has fulfilled any special agreements and conditions that were a part of the building permit.
3. The municipality's Fine Arts Commission or similar authorized body who originally judged whether the structure would be acceptable in the destined area.

Thus, although the builder has a contract with only one party, the owner, he must satisfy the Architect/Engineer who is the official representative of the owner and the building department, which controls all building projects; and sometimes he (with the help of the Architect) must convince another authority such as a fine arts commission that the structure fulfills the intents which persuaded that commission to add its approval for the erection of the structure. The final result will be that the Department of Buildings will agree that the structure is habitable and will grant the owner a Certificate of Occupancy. In most instances the department will grant a Temporary Certificate of Occupancy, which will allow the owner to take over part or all of a building while minor requirements are unfulfilled. Such temporary certificates are written for two or three months and are renewable if final requirements are not achieved.

Even though many contractors help in achieving a Certificate of Occupancy, it is the Architect's responsibility to obtain the temporary and final Certificates of Occupancy from the Department of Buildings. In most large municipalities and cities, these certificates are given by the local Department of Buildings. In some States the inspections and certificates are provided by the county.

32–1   Interim Inspections

Generally, a building or structure cannot obtain final approval of the designer or the building department through only a final one-shot inspection. The Architect/Engineer will make periodic inspections to be sure that the structure is being correctly erected, as will the inspector from the building department.

32–2   Punch Listing

When the builder decides that he has completed a certain area of work in its entirety, he will inspect the area and make a deficiency list for each trade. This is usually called a **punch list**. After the builder has made his **own** punch list for an area he will give copies to subcontractors who are responsible for work he feels is unacceptable, and will reinspect when they advise him that they have completed the remedial work assigned to them. If he feels that all the remedial work he listed is satisfactorily completed, he will formally advise the Architect/Engineer that the area is ready for A/E final inspection and punch listing.

There will be a number of final punch lists for each area, for example, architectural, electrical, plumbing, and HVAC. In addition, in certain areas, such as language laboratories in a school, there will be specialty punch lists for the sound consoles, and where there is a large kitchen or food-preparation area, there will be specialty punch lists made by the Food Service consultant.

When the builder or General Contractor receives these punch lists from the Architect/Engineer, he should check the responsibility for each item or group of items and forward copies (usually two) to each subcontractor involved. The transmittal of these punch lists should always state that the subcontractor should advise the builder *in writing* when his remedial work is completed and reinspection can be made. Notification in writing is important. The builder will be required to inform the Architect/Engineer in writing of these facts, and he should not be forced to make a reinspection until the remedial work is really completed. When the builder has reinspected and has ascertained that all items shown on the A/E punch lists are completed, he will send a formal notification to the A/E to reinspect.

There are times when an Architect will list deficiencies on a punch list
that do not fall within the scope of the contract. In such cases, the builder
will take exception to these items either immediately or when he writes to
inform the A/E that his punch lists are completed. In the case of punch lists
made by the builder or general contractor for his subcontractors, and in the
case where he reinspects after he is advised by his subcontractors that an
A/E punch list has been completed, he must be honest with himself and
with the A/E. He should not advise that a punch list is completed until he
has inspected and concurs that it *is* completed. After all, he accepted a
contract to erect a structure in accordance with documented requirements,
not the least of which was the requirement that the work be accomplished in
accordance with accepted good practice. Thus, if plaster, paint, or other
finished surfaces are not good he should not try to have them accepted by
the Architect. If mechanical/electrical equipment is not correctly installed or
is not operating correctly, he should have these installations corrected even
before he asks for preliminary inspection.

32–3   Guarantees

In general, the contract documents will require that every item and
component of a building or structure be guaranteed for at least one year
from the date of final acceptance by the owner. In addition, there is usually
a 5- or 10-year guarantee requirement for the roofing, 3- to 5-year guarantees
on waterproofing and sealant work, 5-year guarantees on the lamination of
glass, and extended guarantees of 2 to 3 years on certain mechanical and
electrical equipment. In order to close out a project so that these guarantee
periods can officially commence, the written and signed guarantees must be
delivered to the owner (via the Architect) but, more importantly, the
building or structure must be completed **in its entirety** so that the guarantee
periods can commence. Unfortunately, for many subcontractors and equipment manufacturers final completion of a few items and several punch lists
often delays final acceptance by the owner. There are hundreds of cases
each year where the lethargy of certain contractors or subcontractors delays
final acceptance for 6 months or more, which has the result of extending the
guarantees on *every portion* of the contract. There will be no acceptance of
a building or structure unless all the guarantee papers have been delivered.
However, even after these papers are delivered, there will be no official
commencement of any guarantee period unless all facets of the contract are
completed and the owner has made final acceptance of the premises. Thus it
is important to follow up on all deficiency lists and punch lists with vigor.

It is important to note that, even if a building has been officially accepted, there are certain guarantee periods which cannot commence at that
time. One example would be the air-conditioning plant. The contract may
stipulate a guarantee period. However, this guarantee period cannot commence until the equipment is used for an entire air-conditioning season. If

the installation of the chillers of a building is not completed until late September or October, they cannot be accepted, in temperate zones, until the following summer when they can be tested under full heat-load conditions. In such cases the Mechanical consultant will recommend full payment of the HVAC system with the understanding that guarantee periods for chiller equipment will commence after performance acceptance in the next cooling season.

32–4   Delivery of Operating Manuals and Instruction of Operating Personnel

Almost every piece of electrical or mechanical equipment is intricate to the point that operating manuals are necessary. Thus owners and mechanical/electrical specifications require that the supplying subcontractor deliver two or three bound volumes, which contain the operating manuals for all mechanical and electrical equipment. These manuals will not only show how the equipment should be operated, but illustrations of the equipment will refer to part numbers so that replacement parts may be ordered at a later date. Similar operating manuals will be required by a Food-Service specification. These manuals should be formally submitted to the Architect for review by his Mechanical, Electrical, and Food-Service consultants, and for his transmittal to the owner if he and his consultants find the manuals to be in order.

After the Owner has received these manuals and after his operating personnel have had sufficient time to look them over, the contractor must schedule instruction periods that are convenient to the Owner's personnel and to manufacturers' representatives. At these sessions the manufacturers' representatives will meet with the supplying subcontractor and with the owner's maintenance personnel so that they are fully instructed in the operation and maintenance of each piece of equipment. In order that everyone involved is fully protected, it is important that the General Contractor or Construction Management organization list the date and the time of each instruction period and list **all attendees.** In addition, these lists should be signed by each attendee who is a member of the Owner's operating personnel, by their supervisor, and by the representative of the Mechanical/Electrical consultant. The latter **must** be at these training sessions to attest that they were adequate.

These training records are most important! Far too often there is a claim that the session was not complete or that the correct personnel were not given the opportunity to attend. Thus, if the General Contractor or Contract Management organization has acceptance papers signed by the owner's and the designer's representatives, costly additional sessions will not be required. Naturally, if the equipment being demonstrated does not function correctly, another training session will probably be required.

32–5   Certificate of Occupancy

In the beginning of this chapter we noted that the building must be acceptable to the designer, to the Department of Buildings, and (in some instances) to a Fine Arts Commission. Sections 32–1 through 32–4 have listed the procedures necessary to make a building or structure acceptable to the architect and his consultants. If a "Fine Arts" acceptance is necessary, the Department of Buildings will not issue a final Certificate of Occupancy until the owner has received an acceptance letter from the fine arts authority. This will be filed by the Architect when he applies for a final Certificate of Occupancy.

In many buildings, especially high-rise buildings when street-level stores and lower office floors are completed long before the upper floors are completed, decorated, and made habitable, there is a chance to allow the lower floors to be rented. Thus, if the inspector for the Department of Buildings finds that these lower floors can be safely used and that all safety equipment, fire stairs, and exits are in compliance, a request filed by the Architect will often result in the issue of a Temporary Certificate of Occupancy. Of course, this TCO will be financially gainful to the owner. However, in addition, it *may* be beneficial to the contractor in that certain temporary-lighting, temporary-heating, and temporary-elevator men can be removed from the project in the areas that have the TCO. Elevator unions sometimes require a **final** certificate before they will allow their temporary operators to be removed from the project. Regardless, a TCO is most useful to all concerned, and preparations for its issue should be followed with vigor when the time is right. A Final Certificate of Occupancy may often be obtained before all the minor punch lists and remedial work are completed. The important requirements for a final certificate are final completion of the basic structure, completion of all stairs, emergency exits, and fire-alarm systems (which is required for the previously obtained Temporary Certificates of Occupancy), sign-offs by municipal regulatory agencies, and Certificates of Compliance of Design by the Architect and his Structural and Mechanical/Electrical consultants. The Final Certificate of Occupancy may list the occupancy allowed in each area or floor and will always list the per square foot floor-loading allowance as shown by the structural design. When this final certificate has been issued to the Owner, and when remedial work required by all deficiency, omission, and punch lists has been acceptably completed, the architect and owner may be approached with a requisition for final payment.

32–6   Release of Liens and Final Payment for the Entire Project

Final payments were superficially discussed in Chapter 30 where, in Section 30–7, we discussed final payments (less retainage) for subcontractors who had completed their portion of the work. We noted that the Architect and Owner would be reluctant to release payment for 100 percent of any subdivision until a release of lien was received from each subcontractor requesting final payment.

Now we are discussing final payment for *all* contractors, and we must realize that the Architect and the Owner will be reluctant to have these payments made until all contractors and all direct suppliers involved have submitted a release of lien. Thus a wise General Contractor or Construction Management organization will request these releases as soon as possible. Therefore, when all subcontractors and direct suppliers have completed their contracts, the contractor will have all the required papers so that a final requisition can be compiled and passed to the Architect for his approval and forwarding to the Owner. When all the requirements covered up to this point have been acceptably fulfilled, the building can be officially accepted and paid for, and guarantee periods will commence. We have noted, however, that there are times when guarantee periods may have to wait until certain conditions when systems involved may be tested and guarantee periods will start. Conversely, there are times when certain superficial deficiencies will not restrain the owner from taking full "gainful occupancy" of the building, and he will be using most of the equipment. At such times the architect and owner may officially accept the building and allow guarantee periods to commence, and the owner will pay substantially all the contract amounts less a fair allowance for minor remedial work still to be accomplished. However, under general situations, guarantee periods commence at the time of final acceptance and final payment.

32–7   Epilog

The construction industry is one of the first to react to recessions, temporary slowdowns, depressions, or whatever your current politician wishes to call them. As far as the construction man is concerned, whatever you call them, they are periods of reduced work. But, then, certain contractors continue to do some work even in times of financial unrest. The answer is **repeat business** from satisfied clients. Any contractor can receive a contract because of a number of reasons when business is good. However, if he does a good job on **this** contract, he will be considered for the next project that company puts onto the market. At any time, this repeat business is important to a contractor. But when times are lean, it is the company's lifeblood. It behooves the Project Manager, the Project Superintendent, and the entire construction organization, including management, to do a good job on every project. In addition, construction company personnel should make every effort to be friendly with the Architect and the Owner's people so that the Owner and the Architect will wish to have *your* construction company do its next project.

If there are extras (and there will be), these must be diplomatically explained and concisely presented. The Owner and his Architect must be assured that your company is doing its best to give the owner the best service possible, along with the least cost possible. If this idea is continually presented by all construction company personnel, from top to bottom, the client will use your construction company for its next project.

In times when there are many construction projects for your company to choose from it is *desirable* to bring in the "repeat business." However, in lean times, this repeat business may very well be the work that **keeps your company in business!** Every effort should be made to be on good terms with an Owner and his Architect. Most often, this friendly relationship is achieved in the field. Thus the Project Manager and the Construction Superintendent have important roles in the future of their company. It is important that these people try to do their best and, more importantly, show the Owner and his Architect that the company feels that they are "valued." This attitude may very well be the cornerstone of continued company success.

The Project Manager and the Construction Superintendent are charged with the duty of achieving a finished project that is financially successful but that also, very importantly, pleases the Owner or client and his Architect. In the long run, repeat business is the answer. If you wish to confirm this fact, check the ratings of the more successful construction companies and check the amount of repeat business in their yearly work.

Index